D0929502

Science and Technology Concepts–Secondary™

Exploring Plate Tectonics

Student Guide

National Science Resources Center

The National Science Resources Center (NSRC) is operated by the Smithsonian Institution to improve the teaching of science in the nation's schools. The NSRC disseminates information about exemplary teaching resources, develops curriculum materials, and conducts outreach programs of leadership development and technical assistance to help school districts implement inquiry-centered science programs.

Smithsonian Institution

The Smithsonian Institution was created by an act of Congress in 1846 "for the increase and diffusion of knowledge..." This independent federal establishment is the world's largest museum complex and is responsible for public and scholarly activities, exhibitions, and research projects nationwide and overseas. Among the objectives of the Smithsonian is the application of its unique resources to enhance elementary and secondary education.

STC Program™ Project Sponsors

National Science Foundation

Bristol-Meyers Squibb Foundation

Dow Chemical Company

DuPont Company

Hewlett-Packard Company

The Robert Wood Johnson Foundation

Carolina Biological Supply Company

Science and Technology Concepts–Secondary™

Exploring **Plate Tectonics**

Student Guide

The STC Program™

Published by Carolina Biological Supply Company
Burlington, North Carolina

NOTICE This material is based upon work supported by the National Science Foundation under Grant No. ESI-9618091. Any opinions, findings, and conclusions or recommendations expressed in this material are those of the authors and do not necessarily reflect views of the National Science Foundation or the Smithsonian Institution.

This project was supported, in part, by the **National Science Foundation**.
Opinions expressed are those of the authors and not necessarily those of the foundation.

 First Edition 2013
10 9 8 7 6 5 4

ISBN 978-1-4350-0676-8

Published by Carolina Biological Supply Company, 2700 York Road, Burlington, NC 27215.
Call toll free 1-800-334-5551.

1510

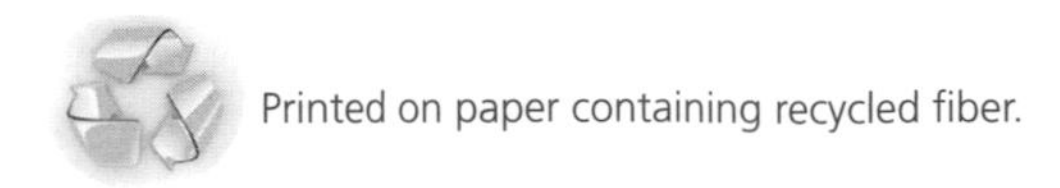

Science and Technology Concepts—Secondary™
Exploring Plate Tectonics

The following revision was based on the STC/MS™ module *Catastrophic Events.*

Developers
Meg Town
Dane J. Toler

Scientific Reviewer
Ian MacGregor
Senior Scientist
National Science Resources Center

Illustrator
John Norton

Developer/Writer
Interactive Whiteboard Activities
Sandy Ledwell, Ed.D

Writers/Editors
Amy Charles
Ian Mark Brooks
Devin Reese

Photo Research
Jane Martin
Devin Reese

National Science Resources Center Staff

Executive Director
Thomas Emrick

Program Specialist/Revision Manager
Elizabeth Klemick

Contractor, Curriculum Research and Development
Devin Reese

Publications Graphics Specialist
Heidi M. Kupke

Carolina Biological Supply Company Staff

Director of Product and Development
Cindy Morgan

Marketing Manager, STC–Secondary™
Jeff Frates

Curriculum Editors
Lauren Eggiman
Gary Metheny

Managing Editor, Curriculum Materials
Cindy Vines Bright

Publications Designers
Trey Foster
Charles Thacker
Weldon D. Washington II
Greg Willette

Science and Technology Concepts for Middle Schools™
Catastrophic Events
Original Publication

Module Development Staff

Developer/Writer
Carol O'Donnell

Science Advisors

Stan Doore, Meteorologist (retired)
National Weather Service
National Oceanic and Atmospheric Administration

Ann Dorr, Earth Science Teacher (retired)
Fairfax County Public Schools, Virginia
Board Member, Minerals Information Institute

Ian MacGregor, Director
Division of Earth Sciences
National Science Foundation

Grant Woodwell, Professor of Geology
Mary Washington College

Thomas Wright, Geologist
National Museum of Natural History
Smithsonian Institution
U.S. Geological Survey (emeritus)

Contributing Writer
Elaine Friebele

Illustrators
John Norton
Max-Karl Winkler

STC/MS™ Project Staff

Principal Investigator
Douglas Lapp, Executive Director, NSRC
Sally Goetz Shuler, Deputy Director, NSRC

Project Director
Kitty Lou Smith

Curriculum Developers
David Marsland
Henry Milne
Carol O'Donnell
Dane J. Toler

Illustration Coordinator
Max-Karl Winkler

Photo Editor
Janice Campion

Graphic Designer
Heidi M. Kupke

STC/MS™ Project Advisors

Judy Barille, Chemistry Teacher, Fairfax County Public Schools, Virginia

Steve Christiansen, Science Instructional Specialist, Montgomery County Public Schools, Maryland

John Collette, Director of Scientific Affairs (retired), DuPont Company

Cristine Creange, Biology Teacher, Fairfax County Public Schools, Virginia

Robert DeHaan, Professor of Physiology, Emory University Medical School

Stan Doore, Meteorologist (retired), National Weather Service, National Oceanic and Atmospheric Administration

Ann Dorr, Earth Science Teacher (retired), Fairfax County Public Schools, Virginia; Board Member, Minerals Information Institute

Yvonne Forsberg, Physiologist, Howard Hughes Medical Center

John Gastineau, Physics Consultant, Vernier Corporation

Patricia A. Hagan, Science Project Specialist, Montgomery County Public Schools, Maryland

Alfred Hall, Staff Associate, Eisenhower Regional Consortium at Appalachian Educational Laboratory

Connie Hames, Geology Teacher, Stafford County Public Schools, Virginia

Jayne Hart, Professor of Biology, George Mason University

Michelle Kipke, Director, Forum on Adolescence, Institute of Medicine

John Layman, Professor Emeritus of Physics, University of Maryland

Thomas Liao, Professor and Chair, Department of Technology and Society, State University of New York at Stony Brook

Ian MacGregor, Director, Division of Earth Sciences, National Science Foundation

Ed Mathews, Physical Science Teacher, Fairfax County Public Schools, Virginia

Ted Maxwell, Geomorphologist, National Air and Space Museum, Smithsonian Institution

Tom O'Haver, Professor of Chemistry/Science Education, University of Maryland

Robert Ridky, Professor of Geology, University of Maryland

Mary Alice Robinson, Science Teacher, Stafford County Public Schools, Virginia

Bob Ryan, Chief Meteorologist, WRC Channel 4, Washington, D.C.

Michael John Tinnesand, Head, K–12 Science, American Chemical Society

Grant Woodwell, Professor of Geology, Mary Washington College

Thomas Wright, Geologist, National Museum of Natural History, Smithsonian Institution; U.S. Geological Survey (emeritus)

Acknowledgments

The National Science Resources Center gratefully acknowledges the following individuals and school systems for their assistance with the national field-testing of *Catastrophic Events:*

East Bay Educational Collaborative, Rhode Island

Site Coordinator
Ronald D. DeFronzo, Science Specialist
East Bay Educational Collaborative
Director, Kits in Teaching Elementary Science, Portsmouth

Michael J. Brennan, Teacher
Portsmouth Middle School, Portsmouth

Mary J. Hayes, Teacher
Thompson Middle School, Newport

Donna Stouber, Teacher
Kickemuit Middle School, Warren

School District of Greenville County
Greenville, South Carolina

Site Coordinator
Toni Enloe, Teaching and Learning Division

Elayne R. Finkelstein, Teacher
League Academy

Robbie L. Higdon, Teacher
League Academy

Mary Helen Maxwell, Teacher
League Academy

Minneapolis Public Schools
Minneapolis, Minnesota

Site Coordinator
James Bickel, Teacher and Instructional Services

Ann Ginis, Teacher
Benjamin Banneker Community School

Michael Madden, Teacher
Ann Sullivan Communication Center

Holly C. Thompson, Teacher
Franklin Middle School

Montgomery County Public Schools
Montgomery County, Maryland

Site Coordinator
Patricia A. Hagan
Middle School Science Specialist

Theresa Manley Sykes,
Science Resource Teacher
White Oak Middle School

School District of Philadelphia
Philadelphia, Pennsylvania

Site Coordinator
Allen Ruby, Research/Curriculum Specialist,
Talent Development Schools,
Center for Social Organization of Schools,
Johns Hopkins University

Deborah Bambino, Teacher
Central East Middle Annex

Jacqueline Dubin, Teacher
Jay Cooke Middle School

Donald L. Rissover, Teacher
Beeber Middle School

Redwood City School District
Redwood City, California

Site Coordinator
Dorothy Patzia, Science Resource Teacher
Bay Area Schools for Excellence in Education
(BASEE)

Anne Renoir, Teacher
Garfield Charter Middle School, Menlo Park

Sandra Robins, Teacher
Hoover Math and Tech Magnet, Redwood City

Bobbie Stumbaugh, Teacher
Selby Lane School, Atherton

Stafford County Public Schools
Stafford County, Virginia

Site Coordinator
Barry Mathson, Science Coordinator

Jan Pierson, Teacher
Gayle Middle School

Winston Ward, Principal
Gayle Middle School

Michael Wondree, Assistant Principal
Gayle Middle School

The NSRC thanks the following individuals for their assistance during the development of *Catastrophic Events*:

Jody Hayob, Geology Professor
Mary Washington College
Fredericksburg, Virginia

Maureen Kerr
Educational Services Manager
National Air and Space Museum
Educational Services
Smithsonian Institution, Washington, D.C.

Fred Klein, Seismologist
U.S. Geological Survey
Menlo Park, California

James F. Luhr, Curator
Global Volcanism Project
National Museum of Natural History
Smithsonian Institution, Washington, D.C.

Steven Mabry, Electronics Engineer
Technology Management Group, Inc.
Dahlgren, Virginia

Amanda May, Teacher
Mountain View Elementary
Haymarket, Virginia

Charles J. Pitts, Electrical Engineer
Science Application International Corporation
McLean, Virginia

Dennis Schatz, Associate Director
Pacific Science Center, Seattle, Washington

Tom Simkin, Curator
National Museum of Natural History
Smithsonian Institution, Washington, D.C.

Rose Steinet, Photo Librarian
Center for Earth and Planetary Studies
National Air and Space Museum
Smithsonian Institution, Washington, D.C.

Penny Sullivan
American Rescue Dog Association
New York, New York

Terry Teays, Manager of Education Group
Origins Education Forum Scientist
Space Telescope Science Institute
Baltimore, Maryland

Tim Watts, Teacher
Chemistry and Marine Science
Courtland High School
Spotsylvania County Public Schools
Spotsylvania, Virginia

The NSRC appreciates the contribution of its
STC/MS project evaluation consultants—

Program Evaluation Research Group (PERG), Lesley College

Sabra Lee
Researcher, PERG

George Hein
Director (retired), PERG

Center for the Study of Testing, Evaluation,
and Education Policy (CSTEEP), Boston College

Joseph Pedulla
Director, CSTEEP

Preface

Community leaders and state and local school officials across the country are recognizing the need to implement science education programs consistent with the National Science Education Standards to attain the important national goal of scientific literacy for all students in the 21st century. The Standards present a bold vision of science education. They identify what students at various levels should know and be able to do. They also emphasize the importance of transforming the science curriculum to enable students to engage actively in scientific inquiry as a way to develop conceptual understanding as well as problem-solving skills.

The development of effective standards-based, inquiry-centered curriculum materials is a key step in achieving scientific literacy. The National Science Resources Center (NSRC) has responded to this challenge through Science and Technology Concepts-Secondary™. Prior to the development of these materials, there were very few science curriculum resources for secondary students that embodied scientific inquiry and hands-on learning. With the publication of STC-Secondary™, schools will have a rich set of curriculum resources to fill this need.

Since its founding in 1985, the NSRC has made many significant contributions to the goal of achieving scientific literacy for all students. In addition to developing Science and Technology Concepts-Elementary™—an inquiry-centered science curriculum for grades K through 6—the NSRC has been active in disseminating information on science teaching resources, preparing school district leaders to spearhead science education reform, and providing technical assistance to school districts. These programs have had a significant impact on science education throughout the country. The transformation of science education is a challenging task that will continue to require the kind of strategic thinking and insistence on excellence that the NSRC has demonstrated in all of its curriculum development and outreach programs. The Smithsonian Institution, our sponsoring organization, takes great pride in the publication of this exciting new science program for secondary students.

Letter to the Students

Dear Student,

The National Science Resources Center's (NSRC) mission is to improve the learning and teaching of science for K-12 students. As an organization of the Smithsonian Institution, the NSRC is dedicated to the establishment of effective science programs for all students. To contribute to that goal, the NSRC has developed and published two comprehensive, research-based science curriculum programs: Science and Technology Concepts-Elementary™ and Science and Technology Concepts-Secondary™.

By using the STC-Secondary™ curriculum materials, we know that you will build an understanding of important concepts in life, earth, and physical sciences; learn critical-thinking skills; and develop positive attitudes toward science and technology. The National Science Education Standards state that all secondary students "...should be provided opportunities to engage in full and partial inquiries.... With an appropriate curriculum and adequate instruction, ... students can develop the skills of investigation and the understanding that scientific inquiry is guided by knowledge, observations, ideas, and questions."

STC-Secondary also addresses the national technology standards published by the International Technology Education Association. Informed by research and guided by standards, the design of the STC-Secondary units addresses four critical goals:

- Use of effective student and teacher assessment strategies to improve learning and teaching
- Integration of literacy into the learning of science by giving students the lens of language to focus and clarify their thinking and activities
- Enhanced learning using new technologies to help students visualize processes and relationships that are normally invisible or difficult to understand
- Incorporation of strategies to actively engage parents to support the learning process

We hope that by using the STC-Secondary curriculum you will expand your interest, curiosity, and understanding about the world around you. We welcome comments from students and teachers about their experiences with the STC-Secondary program materials.

Thomas Emrick
Executive Director
National Science Resources Center

Navigating an STC–Secondary™ Student Guide

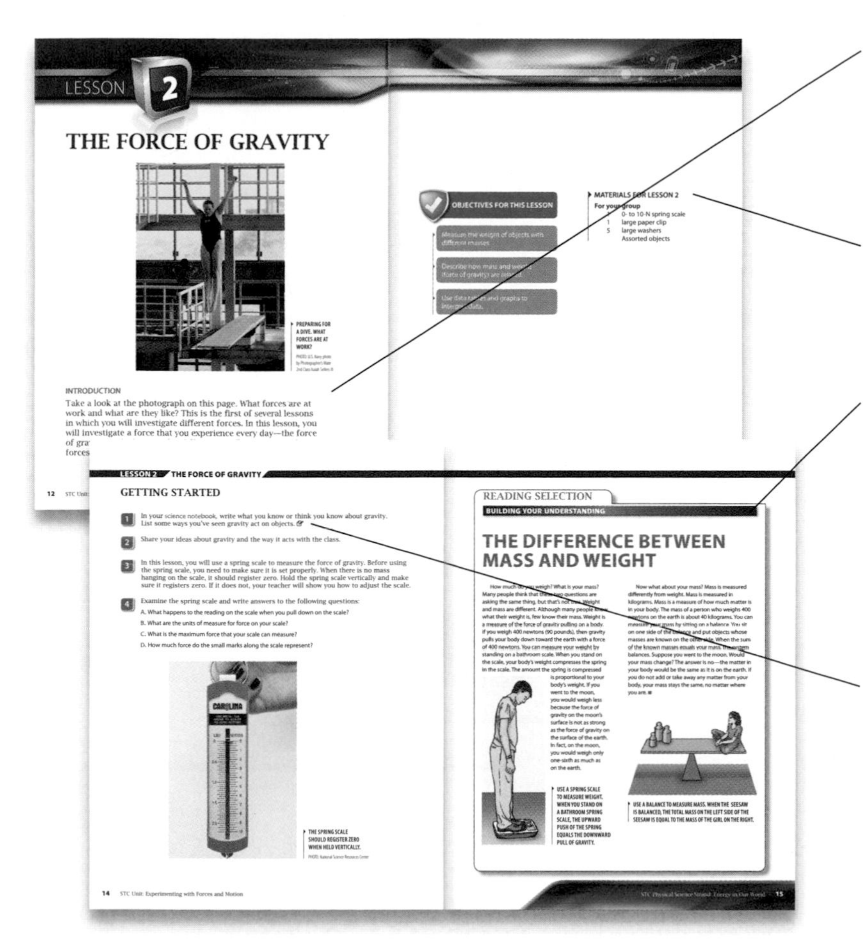

INTRODUCTION
This short paragraph helps get you interested about the upcoming inquiries.

MATERIALS
This helps you get organized and prepare for your inquiries.

READING SELECTION: BUILDING YOUR UNDERSTANDING
These reading selections are part of the lesson, and give you information about the topic or concept you are exploring.

NOTEBOOK ICON
During the course of an inquiry, you'll record data in different ways. This icon lets you know to record in your science notebook. Student sheets are called out when you're to write there. You may go back and forth between your notebook and a student sheet. Watch carefully for the icon throughout the procedure.

SAFETY TIPS
Safety in the science classroom is very important. Tips throughout the student guide will help you to practice safe techniques while conducting investigations. It is very important to read and follow all safety tips.

PROCEDURE
This tells you what to do. Sometimes the steps are very specific, and sometimes they guide you to come up with your own investigation and ways to record data.

REFLECTING ON WHAT YOU'VE DONE
These questions help you think about what you've learned during the lesson's inquiries, apply them to different situations, and generate new questions. Often you'll discuss your ideas with the class.

READING SELECTION: EXTENDING YOUR KNOWLEDGE
These reading selections come after the lesson, and show new ways that the topic or concept you learned about during the lesson can be applied, often in real-world situations.

GLOSSARY
Here you can find scientific terms defined.

INDEX
Locate specific information within the student guide using the index.

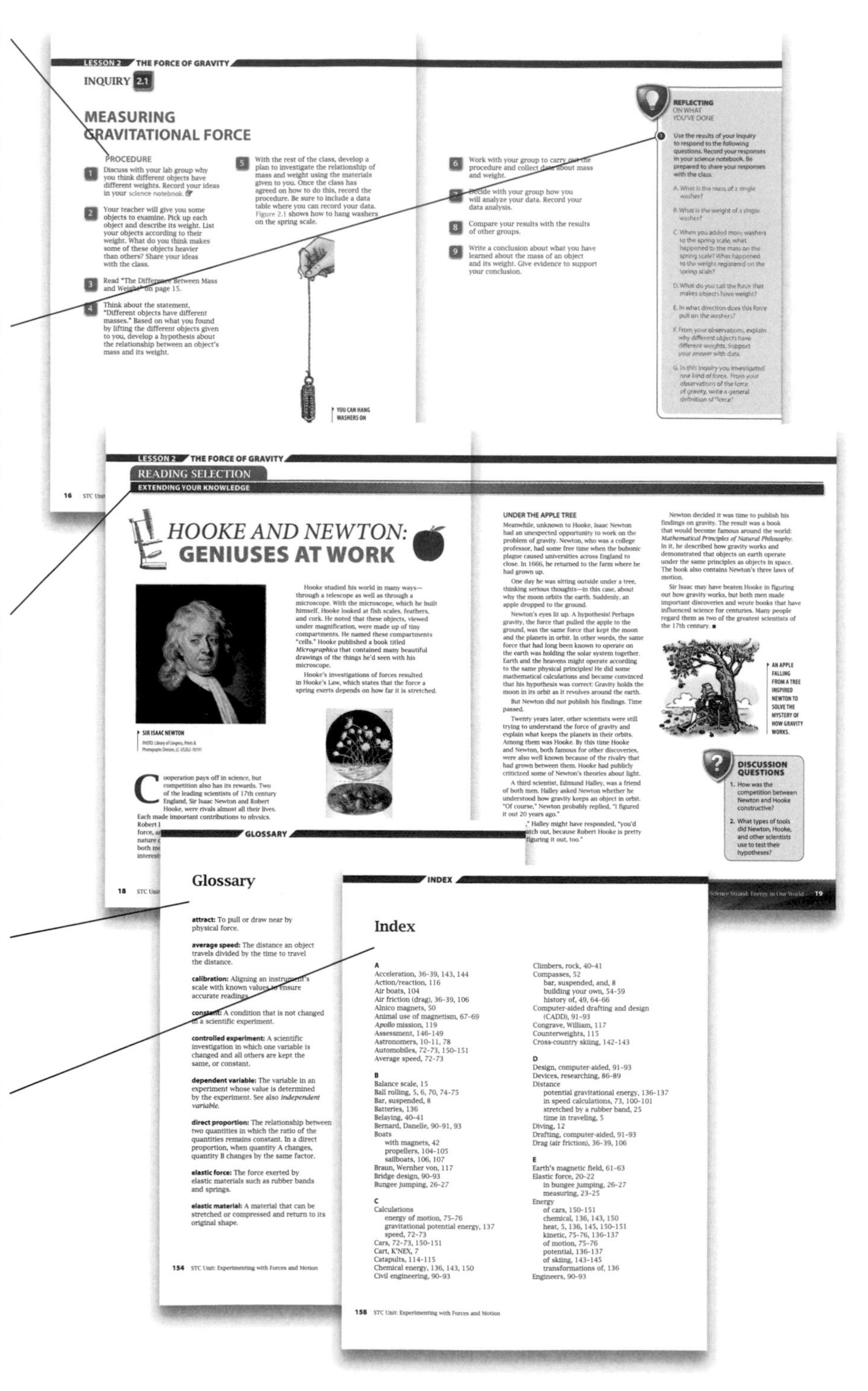

Contents

LESSON 1

INTRODUCING EARTHQUAKES

THIS APARTMENT BUILDING IN CONCEPCIÓN, CHILE, COLLAPSED AS A RESULT OF AN EARTHQUAKE IN FEBRUARY 2010.

PHOTO: U.S. Geological Survey/ photo by Walter D. Mooney, Ph.D.

INTRODUCTION

Did you know that during recorded history, more than 13 million people have died as a result of earthquakes and that about 1 million of those deaths occurred in the 20th century? In one earthquake alone in Turkey in 1999, more than 17,000 people were killed. And, when an earthquake generates a tsunami, the death toll can be even more disastrous. On December 26, 2004, nearly 300,000 people tragically lost their lives in the tsunami that was generated by the Great Sumatra-Andaman earthquake.

In this lesson, you will begin a series of inquiries on earthquakes. You will examine where earthquakes occur most often, and you will use this information to learn more about the earth and why earthquakes happen. What are the greatest risks that individuals face from earthquakes? How might scientists better predict earthquakes? What can society do to reduce the loss of life and property that results from earthquake damage? Let's find out.

OBJECTIVES FOR THIS LESSON

- Brainstorm possible causes and effects of earthquakes and techniques for monitoring and predicting them.
- Use a world map to plot the areas where you believe earthquakes occur.
- View a video of actual earthquakes and discuss the destruction that earthquakes can cause.

MATERIALS FOR LESSON 1

For you

1 copy of Inquiry Master 1.1: Plate Tectonics World Map

For your group

1 sheet of newsprint
1 set of assorted colored markers
1 laminated Plate Tectonics World Map
1 sheet of removable blue dots

GETTING STARTED

1. In your science notebook, write down your ideas about earthquakes. Here are some things to consider:
   A. What do you think causes earthquakes?
   B. What do you think are some of the effects of an earthquake?
   C. What risks do you think an individual might face during or after an earthquake?
   D. How do you think scientists study earthquakes?

2. Share the ideas you wrote down about earthquakes with your group and work together to complete a concept map using the paper and colored markers provided by your teacher.

3. Share your group's concept map and ideas with the class. Your teacher will record your ideas on a class concept map.

RELIEF WORKERS IN HAITI SORT THROUGH RUBBLE AFTER THE 2010 EARTHQUAKE.

PHOTO: Kendra Helmer/U.S. Agency for International Development (USAID)

THINKING ABOUT EARTHQUAKES

PROCEDURE

1. Your teacher will give you a copy of Inquiry Master 1.1: Plate Tectonics World Map. On this paper map, mark the locations where you think earthquakes occur.

2. Now work with your group and the laminated Plate Tectonics World Map and stick the removable blue dots at all the places on the laminated map where the members of your group believe earthquakes occur.

3. Share your laminated map with the class and discuss where you think most earthquakes occur and why.

4. Read "What Is an Earthquake?"

   A. Record in your science notebook any questions you have about earthquakes.

   B. Share your questions with the class. Your teacher will record them on the class brainstorming list. As you work through *Exploring Plate Tectonics*, look for answers to these questions.

READING SELECTION

BUILDING YOUR UNDERSTANDING

WHAT IS AN EARTHQUAKE?

Earthquakes are the shaking and vibrating of the earth. Large and sudden releases of energy cause earthquakes. The energy is released when movement occurs along large "cracks" called faults in the earth's outer layer. More than 90 percent of all recorded earthquakes happen this way. Melted rock on its way to the earth's surface can also cause earthquakes. This usually happens before a volcano erupts. Only about 5 percent of earthquakes are directly related to volcanic activity.

Earthquakes are beyond human control. People know they will happen, and some people are working to reduce the risks associated with them. Seismologists—scientists who study earthquakes—have made great progress in studying the events that come before an earthquake. They attempt to use these events to predict when the next earthquake might occur. The latest technologies use special satellites that monitor the movement of the earth's outer layer, and bore holes that monitor pressure on the rocks. These and other technologies are being used along the San Andreas Fault in California, in Alaska, and in parts of the Himalayas. ■

INQUIRY 

WATCHING EARTHQUAKES HAPPEN

PROCEDURE

1. Read the questions below and then watch the video about earthquakes.

2. Discuss your ideas about the video with the class. In your science notebook, provide evidence from the video to support your answers to these questions:

   A. How does the ground appear to move during an earthquake?

   B. Where do earthquakes usually occur?

   C. What are some destructive, or negative, effects of earthquakes?

   D. Look at Figure 1.1. Are there any constructive, or positive, effects of the earth moving?

REFLECTING ON WHAT YOU'VE DONE

1. Review the concept map that your group completed in Getting Started. Make any additions or changes to the concept map based upon what you have learned in this lesson.

2. Add any new questions you have about earthquakes to the class brainstorming list.

3. Look ahead to Lesson 2, where you will examine earthquake waves.

THE ANDES, A MOUNTAIN SYSTEM ALONG THE WEST COAST OF SOUTH AMERICA, FORMED AS A RESULT OF MANY MOVEMENTS OF THE EARTH OVER A LONG PERIOD OF TIME.

PHOTO: longhorndave/creativecommons.org

FIGURE **1.1**

READING SELECTION

EXTENDING YOUR KNOWLEDGE

Myths About Earthquakes

Throughout history, people have had imaginative and colorful explanations for why earthquakes happen. People in many cultures believed that the earth rested on the back of a massive creature whose movements caused the earth to shake. In Japan, the creature was thought to be a giant catfish; in Mongolia, a giant frog; in China, an ox and a giant tortoise; in India, elephants; and in parts of South America, a whale. The Algonquin Indians of North America thought the earth rested on an immense tortoise.

The people of Siberia, in northern Russia, thought a god called Tuli had a sled so big that it held the earth. They imagined that giant dogs pulled the sled. The dogs had fleas, and when they stopped to scratch, they shook the sled and its cargo—the earth.

IN INDIA, ELEPHANTS SUCH AS THIS ONE AT A TEMPLE, ARE CONSIDERED SACRED. HINDU MYTHOLOGY SAYS THAT THE EARTH WAS HELD UP BY EIGHT STRONG ELEPHANTS, WHO CAUSED EARTHQUAKES BY SHAKING THEIR HEADS.

PHOTO: McKay Savage/creativecommons.org

IN JAPAN, PEOPLE THOUGHT THE EARTH RESTED ON A GIANT CATFISH AND WOULD SHAKE EVERY TIME THE CATFISH MOVED.

READING SELECTION

EXTENDING YOUR KNOWLEDGE

IN 2010, AN EARTHQUAKE IN DICHATO, CHILE, CAUSED A SEA WAVE 10 METERS (33 FEET) HIGH TO WASH OVER PARTS OF THE CITY, LEAVING DESTRUCTION IN ITS WAKE.

PHOTO: U.S. Geological Survey/photo by Walter D. Mooney, Ph.D.

Famous writers and philosophers have sometimes tried to explain earthquakes. For example, the English playwright Shakespeare wrote, "The earth did shake when I was born" (*Henry IV, Part I*). The Greek philosopher Aristotle gave a natural explanation for earthquakes. He thought that atmospheric winds were drawn into the earth's interior. These winds caused fires that swept through underground cavities trying to escape.

In the 1700s, many people believed that earthquakes were a punishment or a warning for those who were not sorry for wrongs they had done. This view was strongly reinforced by the great earthquake that occurred in Lisbon, Portugal, on November 1, 1755. It was All Saints' Day, a holy day, when many people were attending church services. The earthquake was so strong that buildings shook all over Europe. Chandeliers rattled even in parts of the United States. Approximately 70,000 people were killed, mostly from aftershocks (earthquakes that happen after a main earthquake). Buildings collapsed, and a giant sea wave destroyed waterfront areas in Lisbon. Fires burned throughout the city.

Today, most people know that natural events within the earth cause earthquakes. Still, people all over the world fear and wonder about these vibrations. Folklore and legend helped explain these strange and frightful events to people of the past. Technology provides the scientific "stories" that help us understand earthquakes today. ■

DISCUSSION QUESTIONS

1. Why should we believe that scientific "stories" hold more truth than stories about a giant catfish shaking the earth?
2. Do you believe there will be more earthquakes in the world? Explain your reasoning.

LESSON 2

WHEN THE EARTH SHAKES

WHEN A PEBBLE HITS THE WATER, WAVES MOVE OUTWARD IN ALL DIRECTIONS.

PHOTO: Phil Whitehouse/ creativecommons.org

INTRODUCTION

An earthquake occurs when a piece of the earth's crust moves suddenly. This releases a huge amount of energy. The effect of this sudden release of energy is felt later as an earthquake at places hundreds or even thousands of miles away. How does the energy released by this sudden movement make the earth quake at places far away? In this lesson you will investigate this question. You will also investigate how earthquakes can cause major damage to buildings and other structures, and read about how buildings can be made to withstand the destructive forces of earthquakes.

OBJECTIVES FOR THIS LESSON

- Observe the formation and movement of waves in water.
- Conduct a controlled investigation.
- Use a spring to investigate different kinds of waves.
- Relate waves in a spring to earthquake waves.
- Use a spring to model possible damaging effects of earthquake waves.
- Read about how buildings can be designed to withstand earthquakes.

MATERIALS FOR LESSON 2

For you

1 copy of Student Sheet 2.1: Simulating the Motion of Earthquake Waves

For your group

1 spring, with string tied to both ends
1 stopwatch
1 sheet of construction paper
Masking tape

GETTING STARTED

1. With your classmates, review what you learned from the video that you watched during Lesson 1 and discuss the following questions:

   A. How did the ground move during the earthquake?

   B. What were some of the effects of this movement?

2. Look at the plastic box of water on the projector. Make a prediction about how a drop of water that falls into the plastic box will affect the water in it. Discuss your predictions with the class.

3. Watch as a classmate releases a drop of water into the plastic box of water. Discuss with the class what you observe, and propose an explanation for your observations.

4. Think about how this model is related to earthquakes.

HOW WILL THE WATER DROP AFFECT THE WATER BELOW?

INQUIRY 2.1

TESTING THE MOTION OF WAVES

PROCEDURE

1. Look at the spring, the stopwatch, and the masking tape that your teacher has set out.

   A. How might you use these materials to study waves?

   B. What are some ways you could move the spring? Watch as student volunteers discover different ways of moving energy through the spring.

   C. Discuss the different types of waves made by your classmates and drawn by your teacher.

   D. Classify the drawings by your teacher that show different spring movements.

   E. Which type of wave do you think will move the fastest?

   F. Which might do the most damage to buildings on the earth? Discuss your predictions with the class.

2. Before collecting your materials, look at Figure 2.1. Discuss with your teacher how to handle the spring:

   - Do not stretch the spring too far.
   - Do not let go of the spring suddenly. It will tangle.
   - Completely compress the spring before letting go of it.
   - Do not play with the spring in the air. Work only with it on the floor.

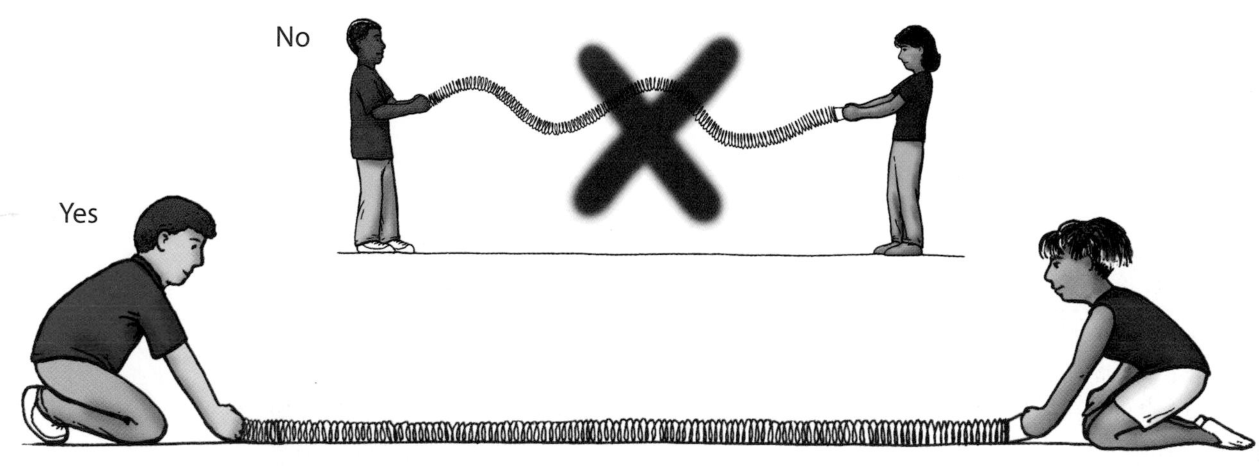

WORK ONLY WITH THE SPRING ON THE FLOOR. DO NOT WORK WITH THE SPRING IN THE AIR.
FIGURE **2.1**

Inquiry 2.1 continued

3 Collect the materials for your group, including Student Sheet 2.1.

A. Be sure the string is securely attached to both ends of the spring (see Figure 2.2).

B. Write the title for your data table. Use one of the formats suggested by your teacher. Be sure to include what you are changing and what you are measuring in the title.

4 Assign roles for your group (timer, recorder, spring mover, and spring holder).

5 Before collecting data, test the motion of the spring by getting it to move in the ways your classmates did earlier (see Figure 2.3).

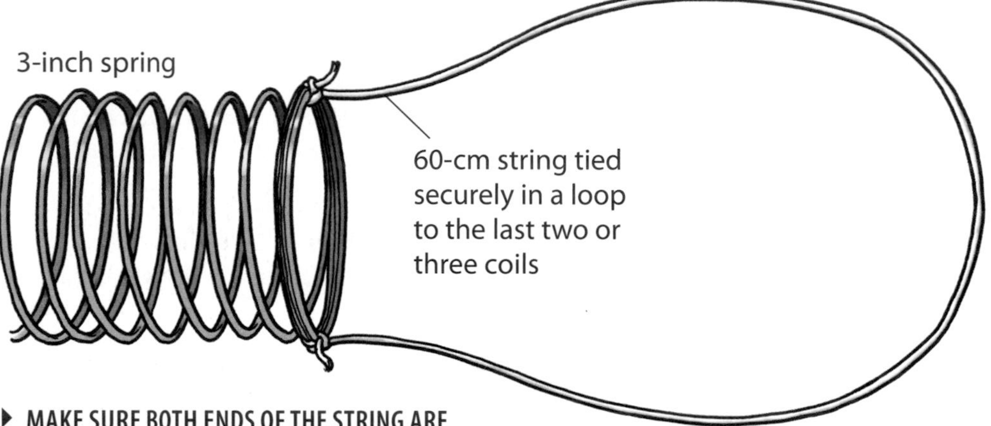

MAKE SURE BOTH ENDS OF THE STRING ARE TIED SECURELY TO THE SPRING. THE LOOP WILL BE A HANDLE TO HOLD THE SPRING.
FIGURE **2.2**

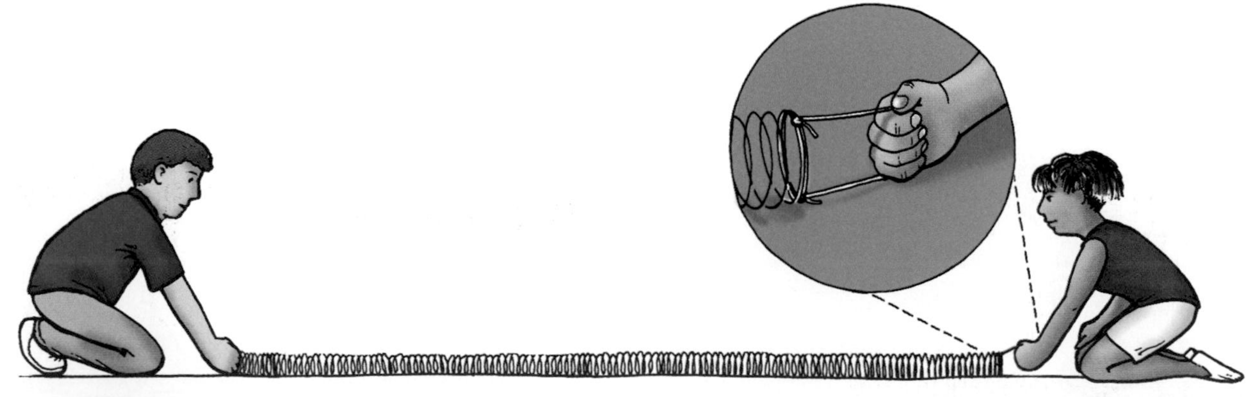

LAY THE SPRING ON THE FLOOR. HOLD THE LOOP OF STRING TIGHTLY IN YOUR HAND.
FIGURE **2.3**

6 Modeling the "push-and-pull" or "P-waves":

A. Use the construction paper to be sure you move the spring the same amount each time (see Figure 2.4). Notice the construction paper is placed lengthwise in front of you.

B. In one quick motion, pull the spring back the length of the construction paper and push the spring forward *one* time). This models a P-wave. Think of it as a "*p*ush-and-*p*ull" wave that moves through the body of the earth.

C. Repeat the motion a few more times to get a good sense of how the spring moves.

D. Sketch the wave in the block of the data table for the "Push and Pull" wave on Student Sheet 2.1.

E. Time the motion of the P-wave as it makes one complete trip, from your hand, up to the hand of your lab partner, and back to your hand again.

F. Repeat the timing of the wave two more times and record your results. Remember to move the spring the same amount each time to be sure this is a fair test.

USE THE CONSTRUCTION PAPER TO GUIDE YOU. KEEP YOUR HAND INSIDE THE BORDERS OF THE PAPER SO THAT YOU MOVE THE SPRING THE SAME DISTANCE EACH TIME.
FIGURE **2.4**

7 Modeling "side-to-side," or "S-waves":

A. Use the construction paper to be sure you move the spring the same amount each time (see Figure 2.5). Notice, you need to turn the paper widthwise.

B. In one quick motion, shake the spring *one* time from side to side. This models an S-wave. Think of it as a "*s*ide-to-*s*ide" wave that moves through the body of the earth.

C. Repeat the motion a few more times to get a good sense of how the spring moves.

D. Sketch the wave in the block of the data table for the "Side to Side" wave on Student Sheet 2.1.

E. Time the motion of the S-wave as it makes one complete trip, from your hand, up to the hand of your lab partner, and back to your hand again.

F. Repeat the timing of the wave two more times and record your results. Remember to move the spring the same amount each time to be sure this is a fair test.

USE A PIECE OF CONSTRUCTION PAPER TO MEASURE THE DISTANCE YOU SHAKE THE SPRING FROM SIDE TO SIDE.
FIGURE **2.5**

Inquiry 2.1 continued

8 Modeling how buildings move at the surface:

A. Place a piece of tape on the end of the spring where the motion first starts (see Figure 2.6).

B. Remember this is a model. This piece of tape will represent a building on the surface of the earth. The spring represents an earthquake wave that moves through the body of the earth. Actual earthquake waves move vertically through the earth (along the spring) from where the earth first moves and arrive at the surface (by your hand).

C. How does the tape move with a P-wave ("push-and-pull")?

D. How does the tape move with an S-wave ("side-to-side")?

E. Discuss your observations and complete Questions 1 and 2 on Student Sheet 2.1.

9 Once "body waves" (P-waves and S-waves) arrive at the surface of the earth, a different kind of wave is made. These new waves are called "surface waves." They move horizontally along the surface of the earth. You will now use your spring to model how houses, fences, and trees move at the surface of the earth due to these surface waves.

A. Put five pieces of tape on five different coils of the spring (see Figure 2.7).

B. Move the spring from side to side.

C. Complete Question 3 on Student Sheet 2.1.

10 Follow your teacher's directions for cleanup. Be sure there are no tangles in the spring.

11 Read "Earthquake Waves and the Transfer of Energy" on page 18.

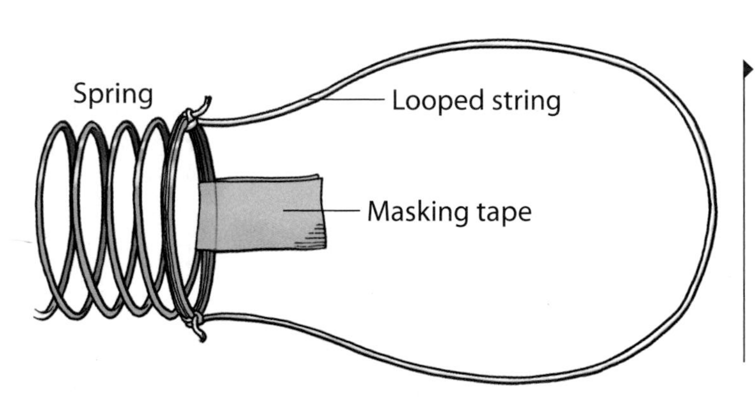

THE TAPE REPRESENTS A STRUCTURE STICKING UP ON THE EARTH'S SURFACE, SUCH AS A BUILDING OR TREE OR HOME (THE MODEL IS TURNED SIDEWAYS). THE SPRING REPRESENTS AN EARTHQUAKE WAVE MOVING THROUGH THE BODY OF THE EARTH, WHICH WOULD ACTUALLY BE COMING UP VERTICALLY FROM BELOW THE EARTH'S SURFACE.
FIGURE **2.6**

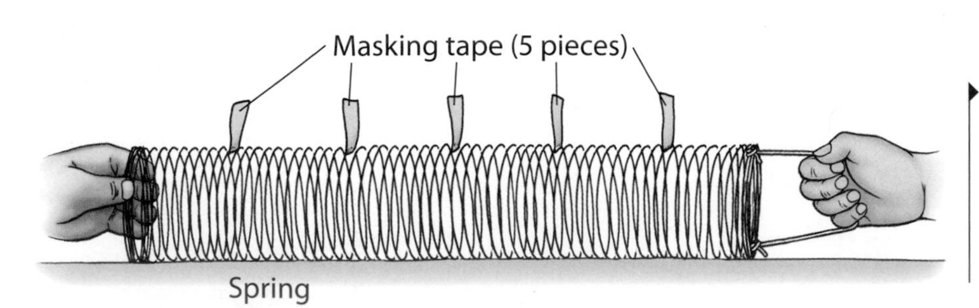

FASTEN THE TAPE ALONG THE OUTSIDE OF THE SPRING. HOW DO SURFACE WAVES MOVE BUILDINGS ON THE EARTH?
FIGURE **2.7**

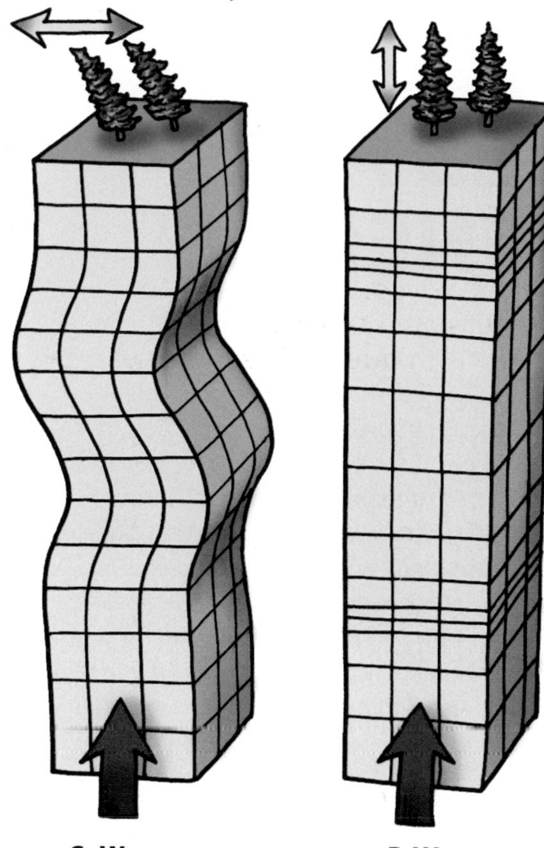

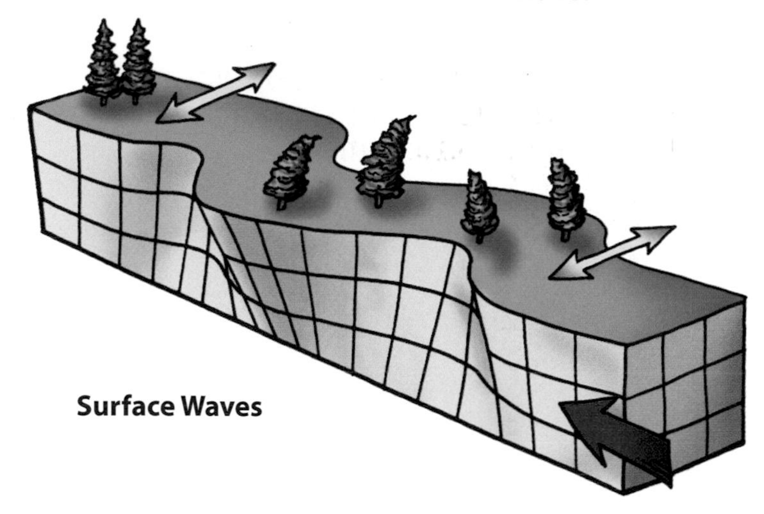

EARTHQUAKE WAVES CAUSE THE GROUND TO MOVE. BODY WAVES MOVE THROUGH THE BODY OF THE EARTH. P-WAVES AND S-WAVES ARE BODY WAVES. SURFACE WAVES MOVE ON OR NEAR THE EARTH'S SURFACE.
FIGURE **2.8**

REFLECTING ON WHAT YOU'VE DONE

1. Describe your observations of each type of wave. Answer the following in your science notebook:

   A. Describe the type of wave created each time you moved the spring.

   B. What was the purpose of the construction paper in this investigation?

   C. Look at your data. Which wave was faster?

   D. If you could not tell which wave was faster, what could have occurred during the investigation to affect your results?

2. Describe your observations of the tape. Answer these questions:

   A. How did each type of wave affect the tape?

   B. Assume the tape represents houses and other structures on the earth's surface. Look at Figure 2.8. How does the ground move with each wave?

3. Read "Designing Earthquake-Resistant Buildings" on pages 19–23 and respond to the following in your science notebook:

   A. Use examples from the reading selection to explain the causes for a building to collapse in an earthquake.

   B. Describe several design features that can be added to a building to make it earthquake-resistant.

4. Look ahead to Lesson 3. Today you learned that energy from an earthquake travels in waves. In Lesson 3, you will investigate how scientists record and study earthquake waves and use the information to pinpoint the origin of an earthquake.

READING SELECTION

BUILDING YOUR UNDERSTANDING

EARTHQUAKE WAVES AND THE TRANSFER OF ENERGY

When a piece of the earth shifts suddenly, earthquake waves move out in all directions, just like sound moves out in all directions when a bell is rung. In fact, a P-wave is just like a sound wave. It pushes particles in the earth closer together and then stretches them apart. That is why we call it a "push-and–pull" wave. Another type of earthquake wave, an S-wave, moves particles in the earth's crust from side to side in a direction perpendicular to the direction of the wave energy. That is why we call it a "side-to-side" wave. In both cases, the energy released by the sudden movement of a piece of the earth's crust moves through the earth's particles, shaking them. But when the wave has passed, the particles come to rest back in their original positions, just as the pieces of tape returned to their original positions in your spring model. This happens because energy simply moves through matter, transferred by waves. We call this a transfer of energy. During an earthquake, things on the earth's surface, such as fence posts, trees, and buildings move up and down or sway side to side. Once the earthquake waves have stopped, the fence posts, trees, and buildings all come to rest in their original positions, unless they have been damaged by the energy in the waves. ■

READING SELECTION

EXTENDING YOUR KNOWLEDGE

Designing EARTHQUAKE-RESISTANT Buildings

"Earthquakes don't kill people. Buildings do." Have you heard that saying? It's true that one of the greatest risks during an earthquake is personal injury from falling debris. What's more, the most populated regions of the world are also places where earthquakes occur most often.

This has happened mainly because many large cities were built long ago in coastal regions near waterways used for shipping. Not until the 1920s did scientists realize that some of these beautiful coastal areas are also the most prone to earthquakes.

People probably will not move away from coastal areas. But today, engineers have learned how to design buildings that are more resistant to earthquakes than those built more than a half-century ago. First, they design a foundation that is

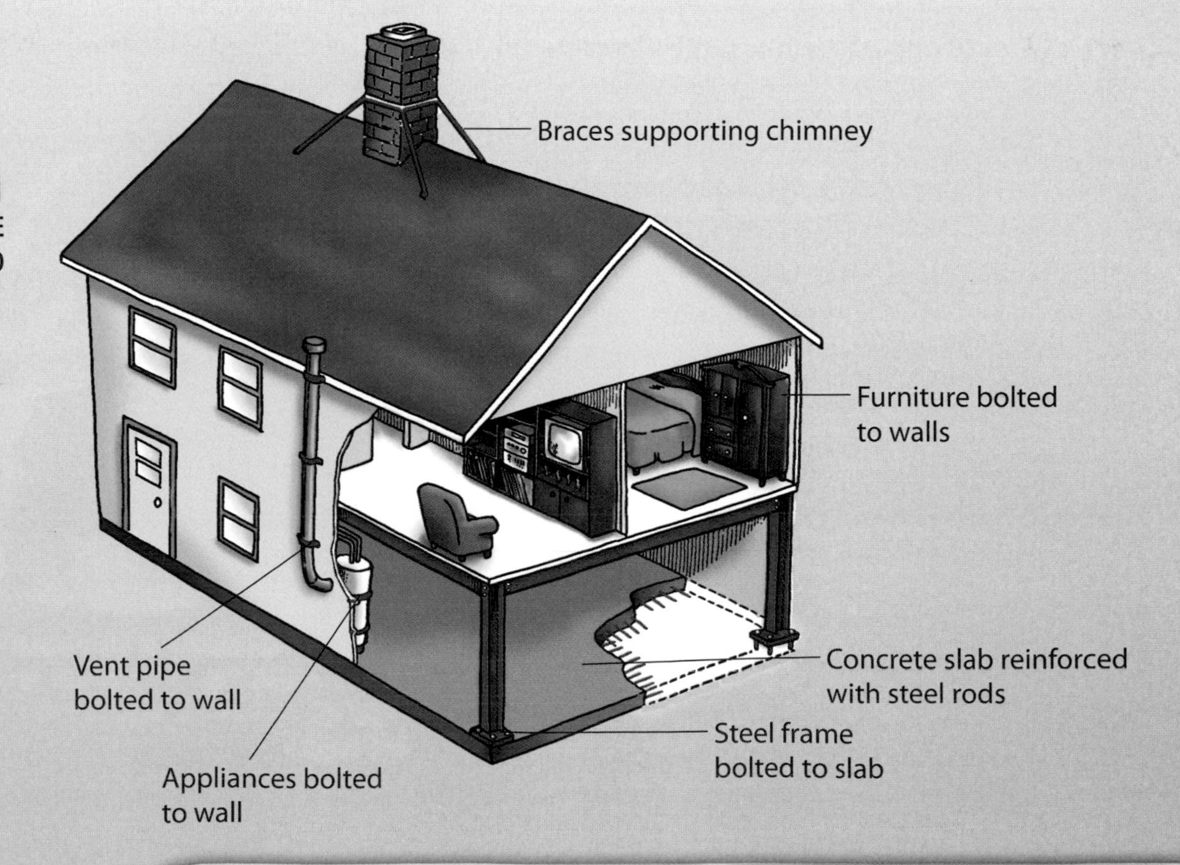

THIS HOUSE SHOWS SOME OF THE THINGS A HOMEOWNER CAN DO TO REDUCE THE RISKS ASSOCIATED WITH GROUND MOVEMENT.

READING SELECTION

EXTENDING YOUR KNOWLEDGE

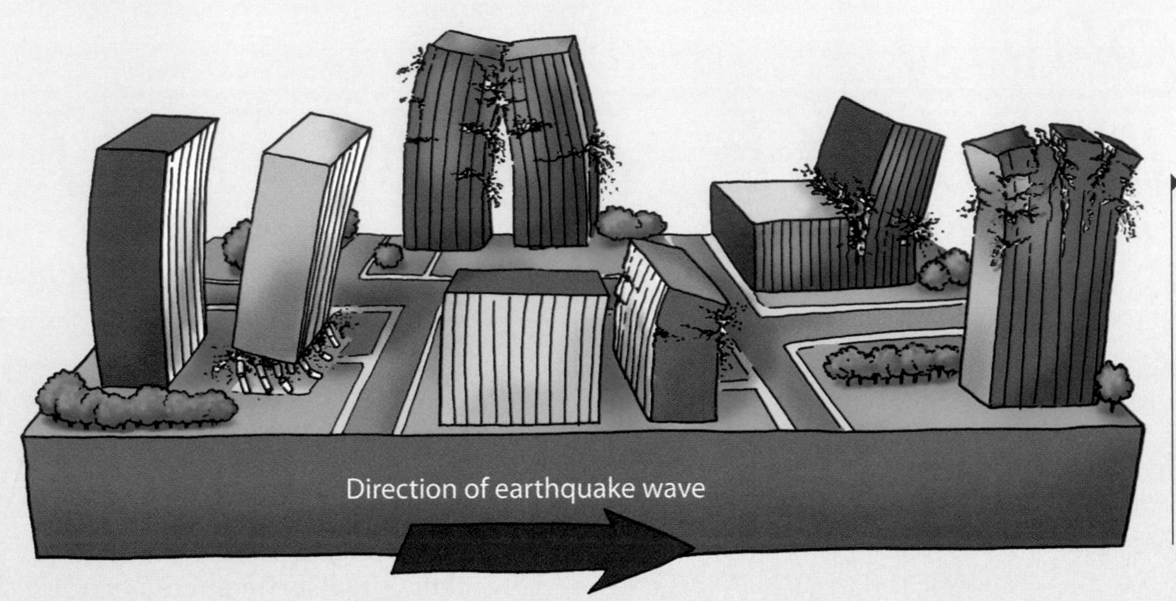

DESIGN MATTERS. THIS ILLUSTRATION SHOWS HOW GROUND MOVEMENT DURING AN EARTHQUAKE AFFECTS BUILDINGS WITH DIFFERENT SHAPES.

firmly connected to solid rock deep in the ground. Second, the building is "tied together." That means the beams and columns that support the structure are strapped together and to the ground with metal, and the floors and roofs are securely fastened to the walls. In California, the masonry walls of many houses at risk are sprayed with liquid concrete and reinforced with steel bracing.

DURING AN EARTHQUAKE IN CALIFORNIA, THE LAYER OF SAND UNDERLYING THIS HIGHWAY BECAME LIQUEFIED, CAUSING THE ROAD SURFACE TO CRACK INTO RUBBLE.

PHOTO: U.S. Geological Survey/photo by S.D. Ellen

WHEN BUILDINGS FAIL

During an earthquake, a building will crack or collapse at places in the structure where there are weak connections. For this reason, in some older neighborhoods that are prone to earthquakes, steel frames are often added to existing structures to strengthen them. Bolting walls to foundations and adding reinforcement beams to the outside of an older home can also help improve its strength.

The size and shape of a building can also affect its resistance to earthquakes. Rectangular, box-shaped buildings are stronger than those of irregular size or shape. This is because different parts of an irregular-shaped building may sway at

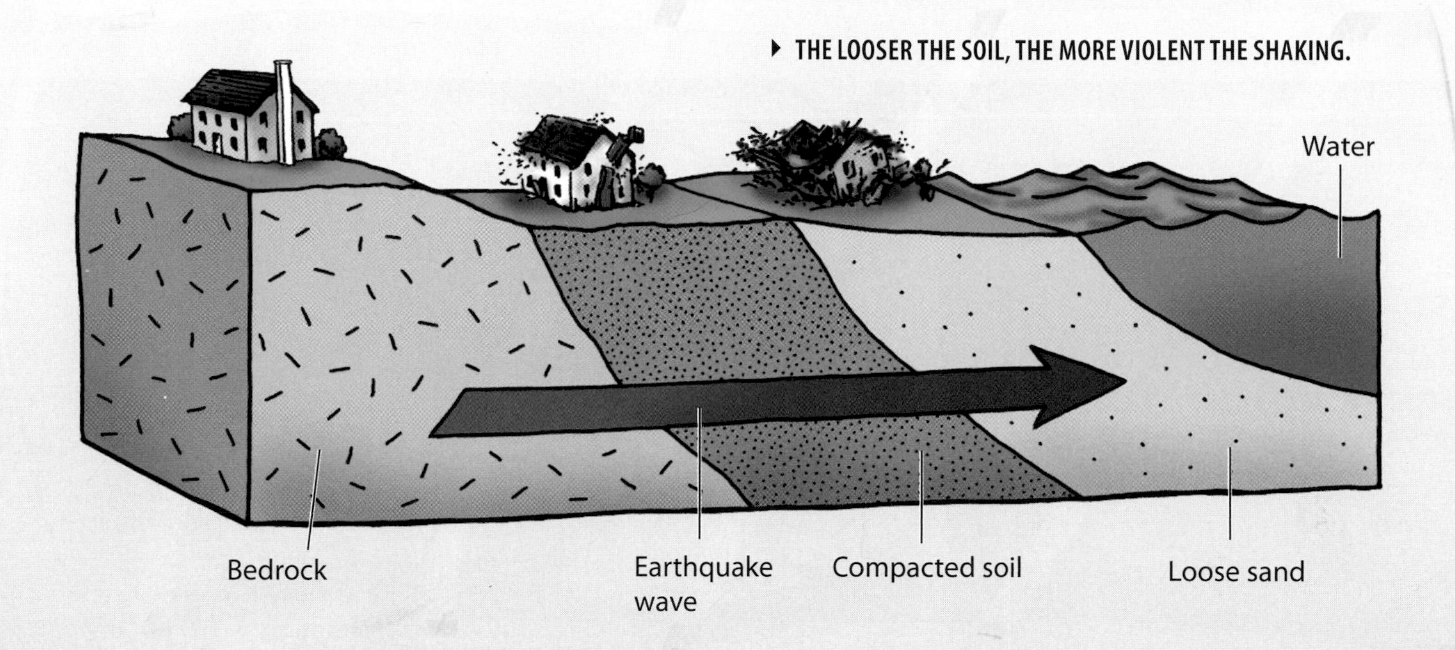

different rates during an earthquake. This puts more stress on the building, which means it is more likely to collapse.

Buildings with open or unsupported first stories are most likely to be damaged during an earthquake. Tall buildings such as skyscrapers must be designed so that a certain amount of swaying or "flexing" can occur, but not so much that they could touch neighboring buildings. A tall building that does not sway slightly will crack and collapse. This is because the stress forces from the earthquake get stronger as they move up the building.

For structures to withstand an earthquake, the ground itself must hold together. Many moist soils—especially those rich in clay—and loose soils, like sand, lose their compactness during an earthquake. When loose or wet soil shakes, parts of the soil rotate. The soil then acts like liquid or gelatin. A building standing on it can collapse as the foundation sinks. Buildings in earthquake-prone areas should stand on solid steel pilings driven deep through a loose or wet soil layer into solid rock deep in the ground.

Constructing earthquake-resistant buildings is cheaper than reinforcing older buildings to make them stronger. Reinforcing entire buildings with supporting frameworks is costly and complicated. Anchoring roofs to supporting masonry is being introduced in some developing countries, but progress is slow.

Even in developing countries, simple buildings made of adobe and similar materials can be made stronger. Techniques include putting bamboo reinforcing strips into the adobe, bracing doorframes and other weak points, and avoiding the use of heavy concrete roofs. ■

READING SELECTION

EXTENDING YOUR KNOWLEDGE

THESE BUILDINGS IN CONCEPCIÓN, CHILE, ARE RAISED 10 METERS (33 FEET) OFF THE GROUND TO WITHSTAND SEA WAVES THAT COULD FOLLOW AN EARTHQUAKE.

PHOTO: U.S. Geological Survey/photo by Walter D. Mooney, Ph.D.

THE BUILDING IN THE FOREGROUND SUFFERED MORE DAMAGE THAN THE BUILDING IN THE BACKGROUND AFTER THE 2010 EARTHQUAKE IN HAITI. WHAT FACTORS COULD EXPLAIN THIS?

PHOTO: U.S. Air Force photo by Master Sgt. Jeremy Lock

DISCUSSION QUESTIONS

1. Which kind of building is more earthquake-resistant: a boxlike skyscraper or a museum built to look like a sculpture? Why?

2. If a house is partially sunken into the ground after an earthquake, what can you infer about the soil under the house?

LESSON 3

RECORDING EARTHQUAKE WAVES

SCIENTIST DR. WALTER MOONEY INSTALLS A SEISMOGRAPH IN A VILLAGE BUILDING TO KEEP TRACK OF AFTERSHOCKS FROM THE 2010 EARTHQUAKE IN HAITI.

PHOTO: U.S. Geological Survey/photo by Walter D. Mooney, Ph.D.

INTRODUCTION

How do scientists know where an earthquake occurred and how strong it was? They use seismographs. Seismographs are instruments that record the vibrations from an earthquake. In this lesson, you will use a model seismograph to record on paper the vibrations you create on a table or other surface. You will then examine a copy of a seismogram recorded during the Alaska earthquake of March 27, 1964. As you study the seismogram, you will discover how scientists record and interpret earthquake waves. Then you will use data recorded from three seismograph stations to model how scientists locate an earthquake's epicenter. This information will help you in Lesson 4, when you plot earthquakes on a world map.

OBJECTIVES FOR THIS LESSON

- Record vibrations using a model seismograph.
- Analyze earthquake wave patterns on an actual seismogram.
- Locate the epicenter of an earthquake using data from three seismograph stations.

MATERIALS FOR LESSON 3

For you

1	copy of Student Sheet 3.1: How Earthquake-Resistant Is Your Home or School?
1	copy of Student Sheet 3.3: Locating the Epicenter of an Earthquake
1	metric ruler
1	3 × 5 in index card

For your group

1	rubber band
1	seismogram
2	drawing compasses

For the class

2	model seismograph stations
2	metric rulers

GETTING STARTED

1. Think back to Lesson 2, and then describe how earthquake waves move. Do they move in only one direction, or do they move outward in circles in all directions? Share your ideas with the class.

2. Brainstorm with your class what you know and want to know about how scientists record and study earthquake waves.

3. Read "A Brief History of Earthquake Detection."

SEISMOMETERS, SENSORS THAT PROVIDE DATA TO A SEISMOGRAPH, WERE INSTALLED IN THE HOTEL MONTANA AFTER AN EARTHQUAKE HIT HAITI IN 2010. WHAT DO YOU THINK SCIENTISTS CAN LEARN BY USING THIS EQUIPMENT?

PHOTO: U.S. Geological Survey/photo by Sue Hough

READING SELECTION

BUILDING YOUR UNDERSTANDING

A BRIEF HISTORY OF EARTHQUAKE DETECTION

Chinese scholar and astronomer Chang Heng invented the first earthquake detector in about 132 AD. This bronze vase (whose replica is shown below) was about 2 meters across, with a domed lid surrounded by eight dragons' heads. Each dragon held a bronze ball in its mouth. Eight bronze toads, with their mouths wide open, were at the base of the vase. The vase contained a heavy column that worked like a pendulum. When the earth shook, the pendulum pushed on a slider, or rod, which dislodged the ball from one of the dragons' mouths. The ball would then fall into one of the toads' open mouths.

As time went on, people invented other instruments to detect earthquakes. But it was not until 1880 that instruments could effectively record the vibrations from earthquakes. At that point, seismology, the study of earthquakes, became a true science. Seismologists began to use mechanical seismographs to detect, record, and measure the vibrations produced by an earthquake. The record made by the seismograph, called a seismogram, was created on a rotating drum. Today, most seismographs are electronic, recording data directly into a computer. ■

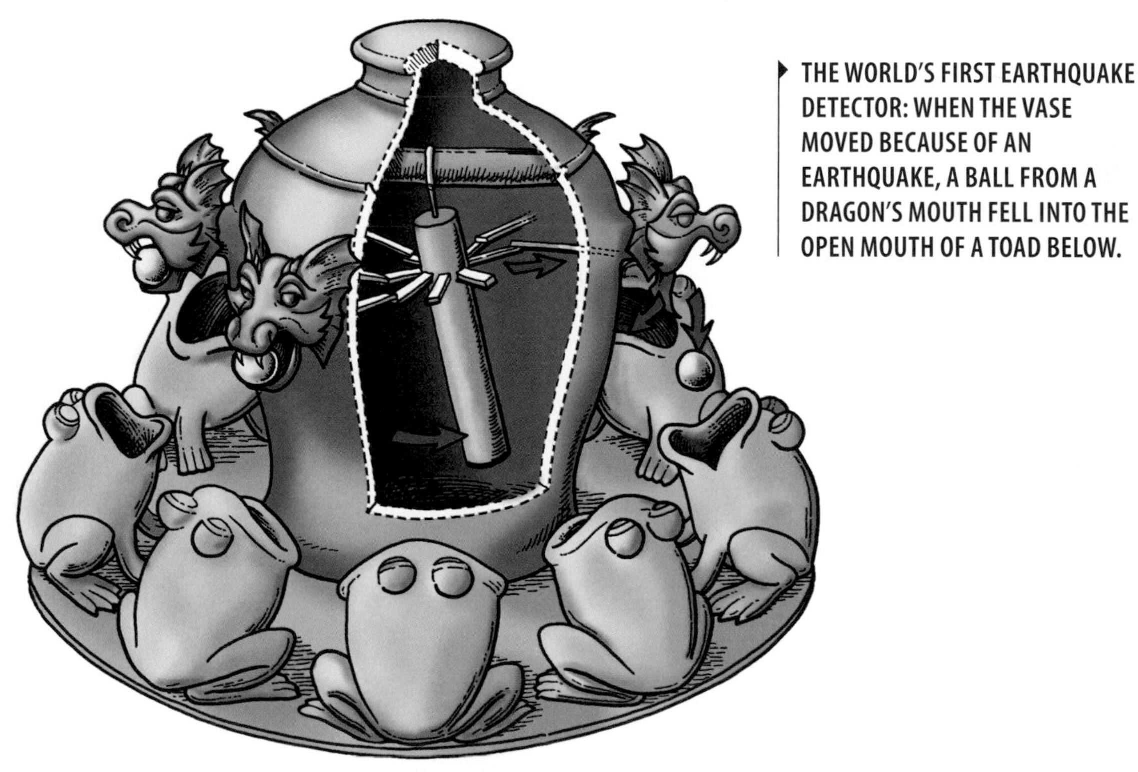

THE WORLD'S FIRST EARTHQUAKE DETECTOR: WHEN THE VASE MOVED BECAUSE OF AN EARTHQUAKE, A BALL FROM A DRAGON'S MOUTH FELL INTO THE OPEN MOUTH OF A TOAD BELOW.

Movement of seismic wave

RECORDING VIBRATIONS

PROCEDURE

1 Look at the model seismograph your teacher shows you. You will use it to record vibrations. Figure 3.1 also shows a model seismograph. How do you think this instrument works? Discuss your ideas with your group. Be prepared to share your group's ideas with the class.

2 As a class, brainstorm ways you could make model earthquakes that this seismograph can record.

3 Discuss with your class how you can test these variables. (Remember that you can test only one of the following variables at a time.)

A. **Direction** How does changing the direction of your pounding (location of your hand) affect the waves that the seismograph records on the paper?

B. **Distance** How does the distance of your pounding from the seismograph affect the waves it records?

C. **Force** How does the strength of your pounding affect the waves it records?

4 With your group, decide which variable in Step 3 you will investigate.

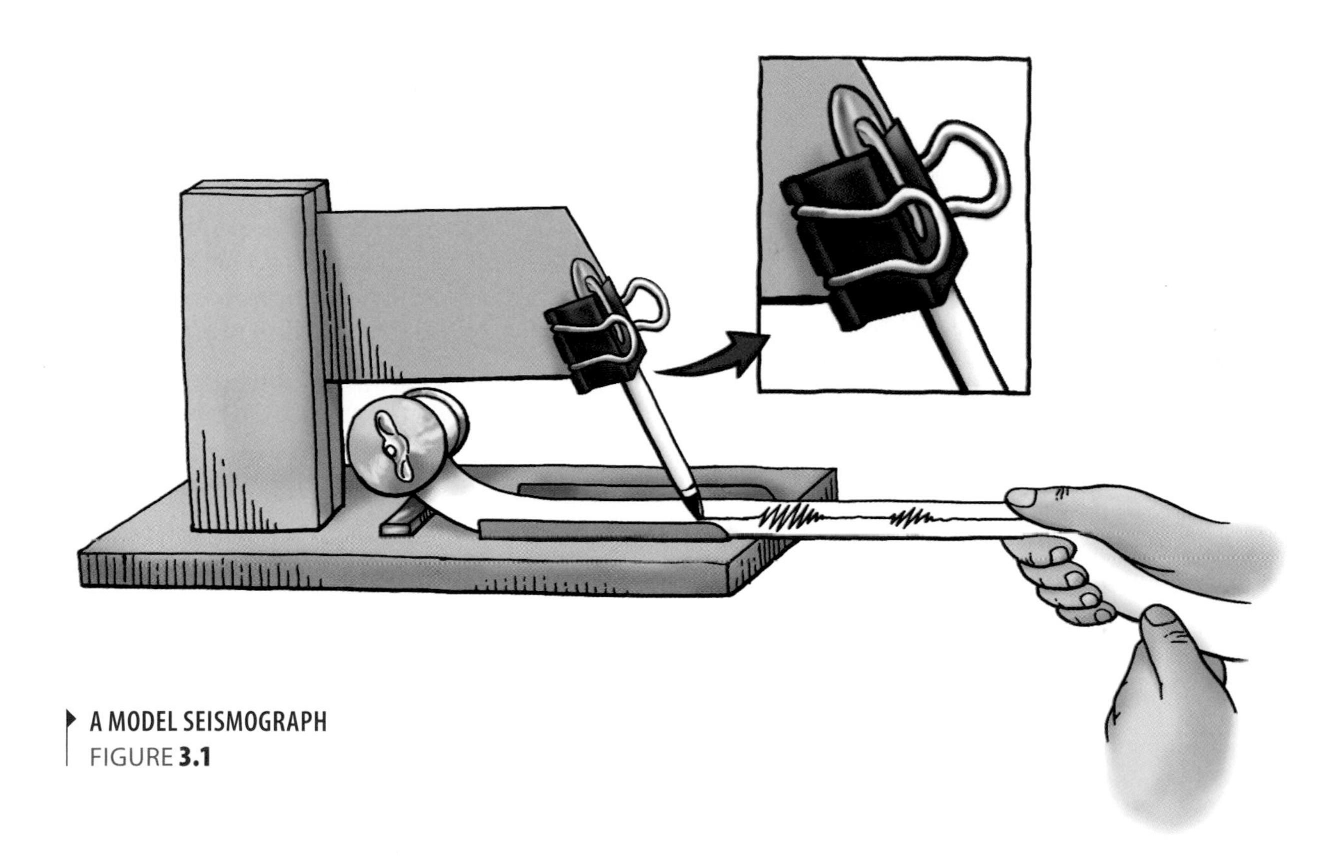

A MODEL SEISMOGRAPH
FIGURE **3.1**

Inquiry 3.1 continued

When directed by your teacher, go to one of the model seismographs set up in the room and follow the directions below for testing your variable using the seismograph. Be prepared to share the results of your investigation with the rest of the class.

DIRECTION

A. Pull the paper strip very slowly through the paper frame. Do not shake the table at all. Is the pen marking the paper? If not, readjust your pen. If it did mark the paper, label your paper strip "Control—No Vibrations." Do not tear off the paper yet.

B. As one person slowly pulls the paper through the frame, have a second person do the following:

1. Pound on the front of the table parallel to the seismograph's arm, as shown in Figure 3.2. Write "Parallel" on the paper.
2. Pound on the side of the table perpendicular to the seismograph's arm, as shown in Figure 3.3. Write "Perpendicular" on the paper.
3. Pound on the surface of the table, as shown in Figure 3.4. Write "Surface" on the paper.

C. Tear the paper strip off the seismograph.

D. Use the paper strip as evidence to support your conclusions as you answer this question: What happened when you changed the direction of your pounding?

E. Roll up the strip and secure it with a rubber band. Write your variable and your group's number on the roll.

POUND ON THE FRONT OF THE TABLE, PARALLEL TO THE SEISMOGRAPH'S ARM.
FIGURE **3.2**

POUND ON THE SIDE OF THE TABLE, PERPENDICULAR TO THE SEISMOGRAPH'S ARM.
FIGURE **3.3**

POUND ON THE SURFACE OF THE TABLE.
FIGURE **3.4**

Inquiry 3.1 *continued*

DISTANCE

A. Pull the paper strip very slowly through the paper frame. Do not shake the table at all. Is the pen marking the paper? If not, readjust your pen. If it did mark the paper, label your paper strip "Control—No Vibrations." Do not tear off the paper yet.

B. Test the variable of distance. With each test, change the distance your seismograph is from the pounding, but keep the direction of your pounding constant. For example, do the following:

1. Pound close to the seismograph. Pound the side of the table that is perpendicular to the seismograph's arm, as shown in Figure 3.3, while your partner pulls the paper through the frame. Measure the distance. Label the paper—for example, "Close to the seismograph—10 cm."
2. Move the seismograph far from your hand, approximately 30 to 40 cm, as shown in Figure 3.5. (If you are using a copier box or plastic box, your distances may be less than 30 cm.) Measure the distance. Remember to pound in the same way and from the same direction as you did when you pounded close to the model. Label the marks on your paper.

C. Tear the paper strip off the seismograph.

D. Use the paper strip as evidence to support your conclusions as you answer this question: What happened when you changed the distance of pounding from the seismograph?

E. Roll up the strip and secure it with a rubber band. Write your variable and your group's number on the roll.

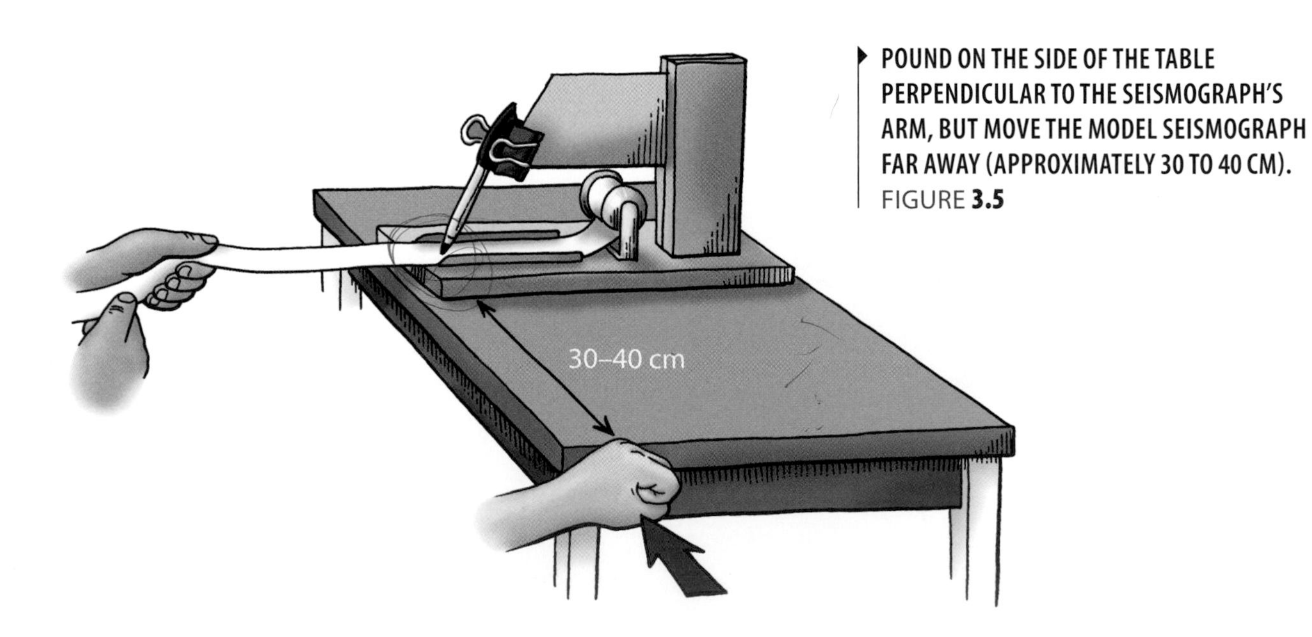

POUND ON THE SIDE OF THE TABLE PERPENDICULAR TO THE SEISMOGRAPH'S ARM, BUT MOVE THE MODEL SEISMOGRAPH FAR AWAY (APPROXIMATELY 30 TO 40 CM).
FIGURE **3.5**

FORCE

A. Pull the paper strip very slowly through the paper frame. Do not shake the table at all. Is the pen marking the paper? If not, readjust your pen. If it did mark the paper, label your paper strip "Control—No Force." Do not tear off the paper yet.

B. Now test the variable of force. Remember to keep the distance and the direction constant. Do the following:

   1. Very gently, pound the side of the table close to and perpendicular to the seismograph's arm. (See Figure 3.3.) Pull the paper through as you pound gently. Mark your paper strip "Small Force."

   2. Pound much harder this time. Make sure you pound from the same direction so that you are only changing one variable at a time. Mark your paper strip "Large Force."

C. Tear the paper strip off the seismograph.

D. Use the paper strip as evidence to support your conclusions as you answer this question: What happened when you changed the force of the pounding?

E. Roll up the strip and secure it with a rubber band. Write your variable and your group's number on the roll.

6 Follow your teacher's directions for sharing with the class what you have learned about the seismograph.

7 Apply your class's findings by answering these questions:

A. How do you think a real seismograph works?

B. How do you think a real seismograph differs from this model seismograph?

C. Why is there always more damage to structures that are built close to the source of an earthquake than to structures that are farther away?

D. Why do you think it is important for scientists to record earthquake vibrations?

READING SELECTION

BUILDING YOUR UNDERSTANDING

THE ALASKA EARTHQUAKE OF 1964

You now know that seismographs record the vibrations of earthquakes, and a seismogram is the record of those vibrations. The seismogram you will use in this inquiry was recorded in Bellingham, Washington, during the earthquake in Prince William Sound, near Valdez, Alaska, on March 27, 1964. This earthquake was one of the most violent ever recorded. The main shock (or earthquake) triggered 1000 aftershocks. These aftershocks occurred in a zone 1000 kilometers long. The ground near the southwestern end of Montague Island, off the coast of Alaska, rose almost 15 meters in the air. Many buildings were nearly split in half.

THIS SCHOOL SPLIT IN HALF WHEN SOME OF THE SOIL UNDER IT MOVED DOWNSLOPE DURING THE EARTHQUAKE IN PRINCE WILLIAM SOUND, ALASKA.

PHOTO: NOAA/National Geophysical Data Center

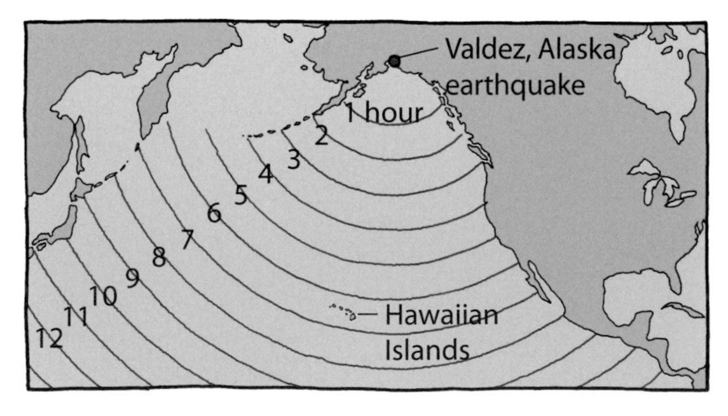

TSUNAMIS CAN BE CAUSED BY EARTHQUAKES. A TSUNAMI CAN TRAVEL FOR THOUSANDS OF MILES. MORE THAN SIX HOURS AFTER THE EARTHQUAKE OCCURRED NEAR VALDEZ, ALASKA, A HUGE TSUNAMI REACHED HAWAII.

Landslides and tsunamis resulted from the earthquake. One tsunami caused flooding in basements in houses as far away as Hawaii. Despite the great strength of the Alaska earthquake, only 122 people died, which is a relatively low number for such a large earthquake. Why? The area in which the earthquake occurred was not highly populated. The tide was low, so the tsunami's effects were not as great. And schools and many businesses were closed because it was Good Friday, a religious holiday. ■

READING A SEISMOGRAM

PROCEDURE

1. Read "The Alaska Earthquake of 1964."

2. In this inquiry, you will use a seismogram recorded in Bellingham, Washington, during the 1964 earthquake in Prince William Sound, Alaska. With your class, find these areas on the geopolitical globe.

3. Look at Figure 3.6 and the transparency your teacher shows you. The transparency explains the times listed on a seismogram. Use the illustration and transparency to answer these questions with your class:

   A. What do the numbers 0858 on the seismogram represent?

   B. What does each mark on a line represent?

   C. How long did it take for the seismogram to make one revolution around the drum? How do you know?

   D. When did the first P-wave in the illustration arrive at the station?

   E. When did the first S-wave arrive at the station?

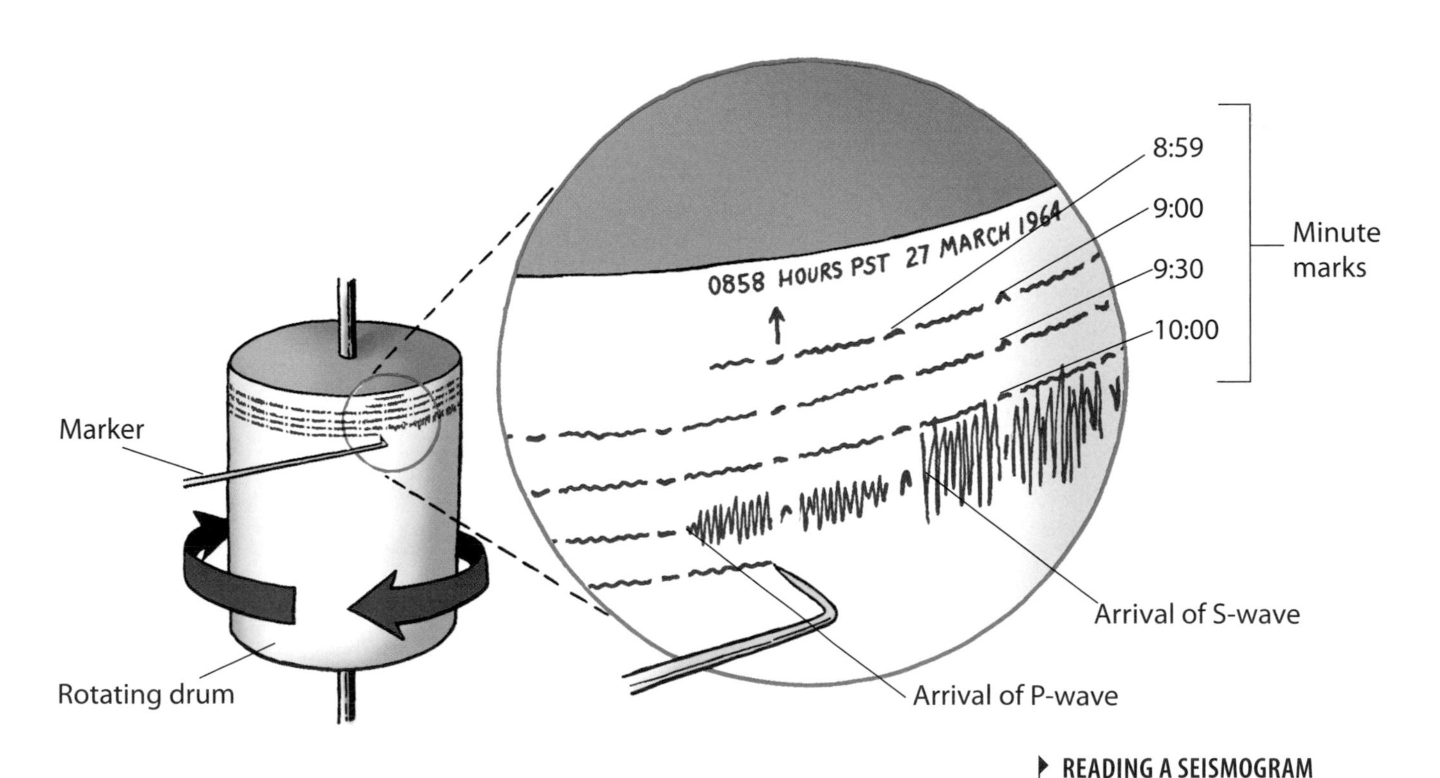

READING A SEISMOGRAM
FIGURE 3.6

Inquiry 3.2 continued

4 Get one seismogram for your group and follow the directions in the following steps. This is a copy of the actual seismogram recorded in Bellingham. Spread out the seismogram on your desk or table. Do not make any pencil or pen marks on it.

5 Look at the Bellingham seismogram and discuss it with your group.

A. How is it different from the seismogram shown in Figure 3.6?

B. How is it the same?

C. Why do you think this is so?

6 Find the spot on the seismogram where the main shock of the Alaska earthquake reached Bellingham. How can you tell? Talk about it with your group.

7 Using the minute marks at the top of the seismogram, determine at what time the first earthquake wave from the Alaska earthquake reached the seismograph station in Bellingham, Washington. Remember:

A. Each mark represents one minute.

B. Each line represents 30 minutes, or 1/2 hour.

C. Use your ruler if needed.

D. Estimate to the nearest minute.

8 In your science notebook, record and label the time the earthquake wave (a P-wave) arrived in Bellingham. Remember this is military time and is recorded on a 24-hour clock. What time would it be in "A.M." or "P.M."?

9 The earthquake occurred in Prince William Sound, Alaska, at 7:36 P.M. (19:36 military time). Using the time you recorded in Procedure Step 8, do the following:

- Calculate how long it took the first earthquake wave to reach Bellingham, Washington.
- Record and label this time.

10 Look at the bottom half of the seismogram. Look at the date. What do you think the waves in this bottom half represent? Discuss this with your group.

11 Find the point on the seismogram that you think represents the first aftershock recorded on March 28, 1964.

A. Record and label this time.

B. Count the number of aftershocks you see altogether on the second day. How many aftershocks do you see altogether? Record the number of aftershocks your group sees.

12. Think back to the marks you made on the paper strip during Inquiry 3.1. Discuss the following questions:

    A. How are they like the marks on this seismogram?

    B. How are they different?

13. Look at the earthquake recorded on the second day that has the P- and S-waves marked. Think back to Lesson 2. What does the position of the P-wave and S-wave on the seismogram tell you about each wave? Discuss this with your group.

14. The arrows on the P- and S-waves indicate when the P-wave first arrived and when the S-wave first arrived. Determine and record the difference in the arrival times of the two waves.

15. The time difference between the P-wave arriving and the S-wave arriving at the seismograph station is called "lag time." Write a working definition of lag time. Be sure to explain how to calculate lag time in your definition. You will use this concept in Inquiry 3.3.

16. What did you learn about reading a seismogram? Record your ideas.

Inquiry 3.2 continued

17 The magnitude of the Alaska earthquake was 9.2, which is higher than that of most recorded earthquakes. Knowing this, answer these questions:

A. Do you think seismographs all over the world, or only those near Alaska, were able to record the Alaska earthquake?

B. In what ways might the seismograms recorded in other parts of the world look different from the one recorded in Bellingham?

18 Look at one of the aftershocks on March 28, and notice how the wave pattern changes.

A. What do the two wave shapes represent?

B. Draw a conceptual graph. A conceptual graph has no numbers; it shows relative shapes and positions of lines.

1. Label the x-axis with "Distance Wave Traveled."
2. Label the y-axis with "Travel Time."
3. Now draw two lines on the graph. Make one line represent the P-wave and the second line represent the S-wave as they travel out from the earthquake. Remember the P-wave travels faster. This is called a "time-distance" graph.

19 Return the seismogram in the same condition in which you received it.

READING SELECTION

BUILDING YOUR UNDERSTANDING

FINDING THE EPICENTER: THE TORTOISE AND THE HARE

Seismographs are located all over the world. A seismograph station in Montana can pick up an earthquake occurring in California. If the earthquake is really strong, seismograph stations all over the world can also record this same earthquake. Why? Because earthquake waves that travel through the body of the earth move outward in all directions.

The point where the earthquake occurs is called the focus, and it can be shallow or deep in the earth. The point on the earth's surface directly above the focus is called the earthquake's epicenter, as shown in the illustration. This is usually the place that you read or hear about in the news when there has been an earthquake.

Damage from one earthquake can occur in many different places. Seismologists use special math calculations to pinpoint the exact location of the earthquake's epicenter. First, they use the arrival times of P- and S-waves, as shown on several seismograms. Then they plot those times on a special graph called a time-distance graph.

P-waves always travel at the same average speed. S-waves have a constant speed, too, only slower. Scientists can graph the speeds of both waves on a special graph. (See Figure 3.7 on page 41.) Like the tortoise in the story "The Tortoise and the Hare," an S-wave always travels more slowly than a P-wave. Although both animals in the story leave the starting line at the same time, the distance between the two becomes greater and greater the farther they are from the starting line (until the hare takes a break). By computing how many minutes apart the two animals are from each other at any one point in the race, you could calculate the distance they have traveled (how far they are from the starting line) using a time-distance graph.

In much the same way, by knowing when each wave arrives at the seismograph station and subtracting the difference, seismologists can determine how far away the earthquake's epicenter is from their station. The greater the difference in time between the P- and S-waves' arrival, the further the seismograph station is from the epicenter.

Is that enough information to pinpoint the exact location of the earthquake? No, because earthquake waves do not follow one path. They move outward in all directions in a circle around the epicenter. With data from a single station, the seismologist knows only that the earthquake could have happened anywhere in a circle around that station. With data from two seismograph stations, however, seismologists can narrow the epicenter location down to two places—that is, the two points at which the circles cross each other. But with data from three stations, they determine the exact point at which all three circles intersect, as shown on the map below. This point is the epicenter of the earthquake. ■

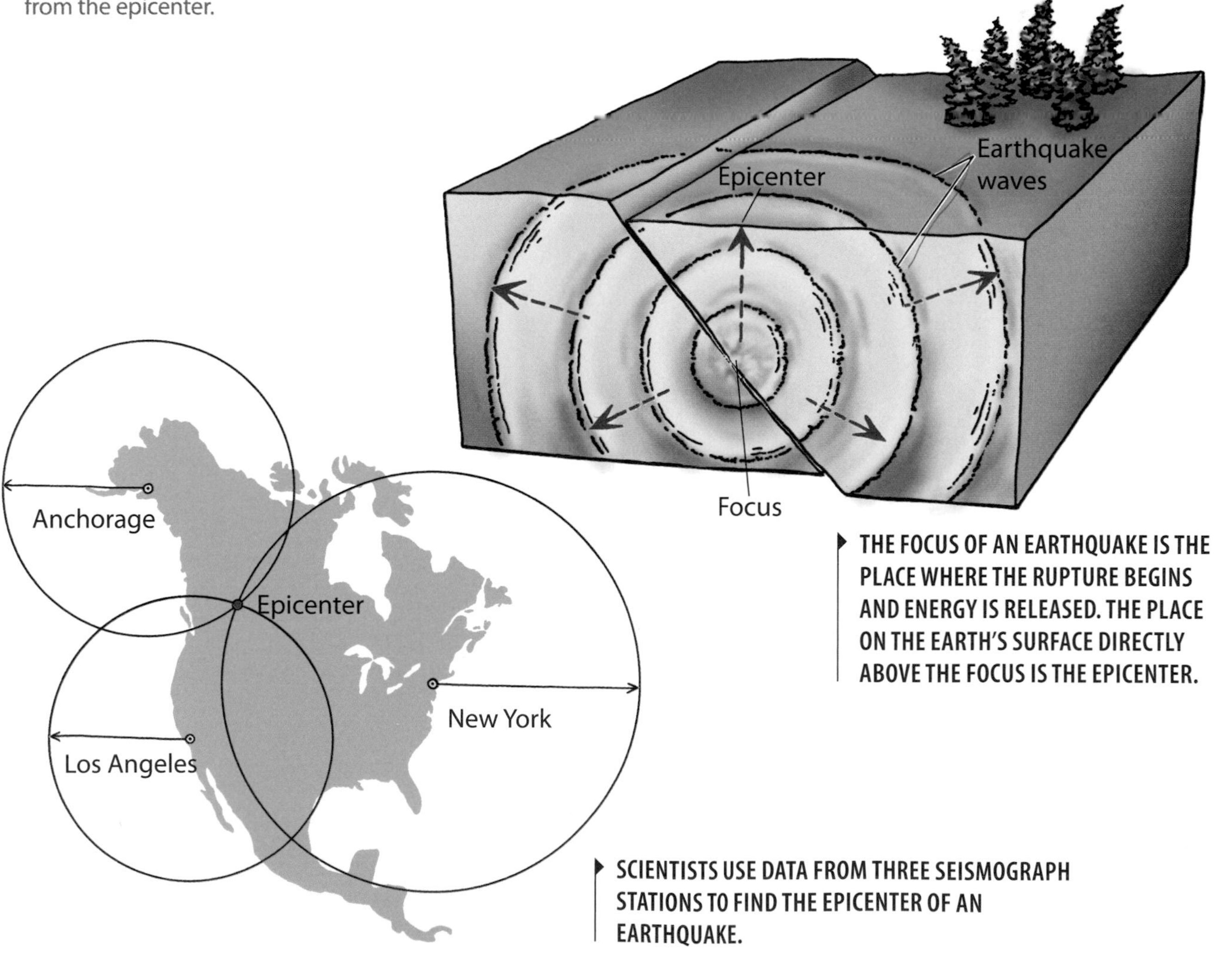

THE FOCUS OF AN EARTHQUAKE IS THE PLACE WHERE THE RUPTURE BEGINS AND ENERGY IS RELEASED. THE PLACE ON THE EARTH'S SURFACE DIRECTLY ABOVE THE FOCUS IS THE EPICENTER.

SCIENTISTS USE DATA FROM THREE SEISMOGRAPH STATIONS TO FIND THE EPICENTER OF AN EARTHQUAKE.

INQUIRY 3.3

LOCATING THE EPICENTER OF AN EARTHQUAKE

PROCEDURE

1. Read "Finding the Epicenter: The Tortoise and the Hare," on pages 38–39.

2. Look again at your data from Inquiry 3.2. Do you think that knowing that the earthquake was four minutes away from Bellingham is enough information to enable you to determine the exact location of the earthquake?

3. Look at the graph in Figure 3.7. To learn how to use the graph, answer these questions:

   A. If it took four minutes for the first P-wave to arrive at the seismograph station, how far away is the earthquake's epicenter? (Hint: Look on the P-wave curve.)

   B. If the seismograph station were located 2500 km from the earthquake's epicenter, how long would it take the P-wave to arrive?

   C. How long would it take the S-wave to travel 2500 km and reach the seismograph station?

   D. Using 2500 km, which wave traveled faster: the P-wave or the S-wave?

   E. At 2500 km, how many minutes elapsed between the time the P-wave and S-wave arrived at the station?

   F. Considering the whole graph, how does minutes elapsed between the P-wave and S-wave arrival relate to distance from the earthquake's epicenter?

4. Pick up your group's drawing compasses, an index card, a ruler, and a copy of Student Sheet 3.3: Locating the Epicenter of an Earthquake.

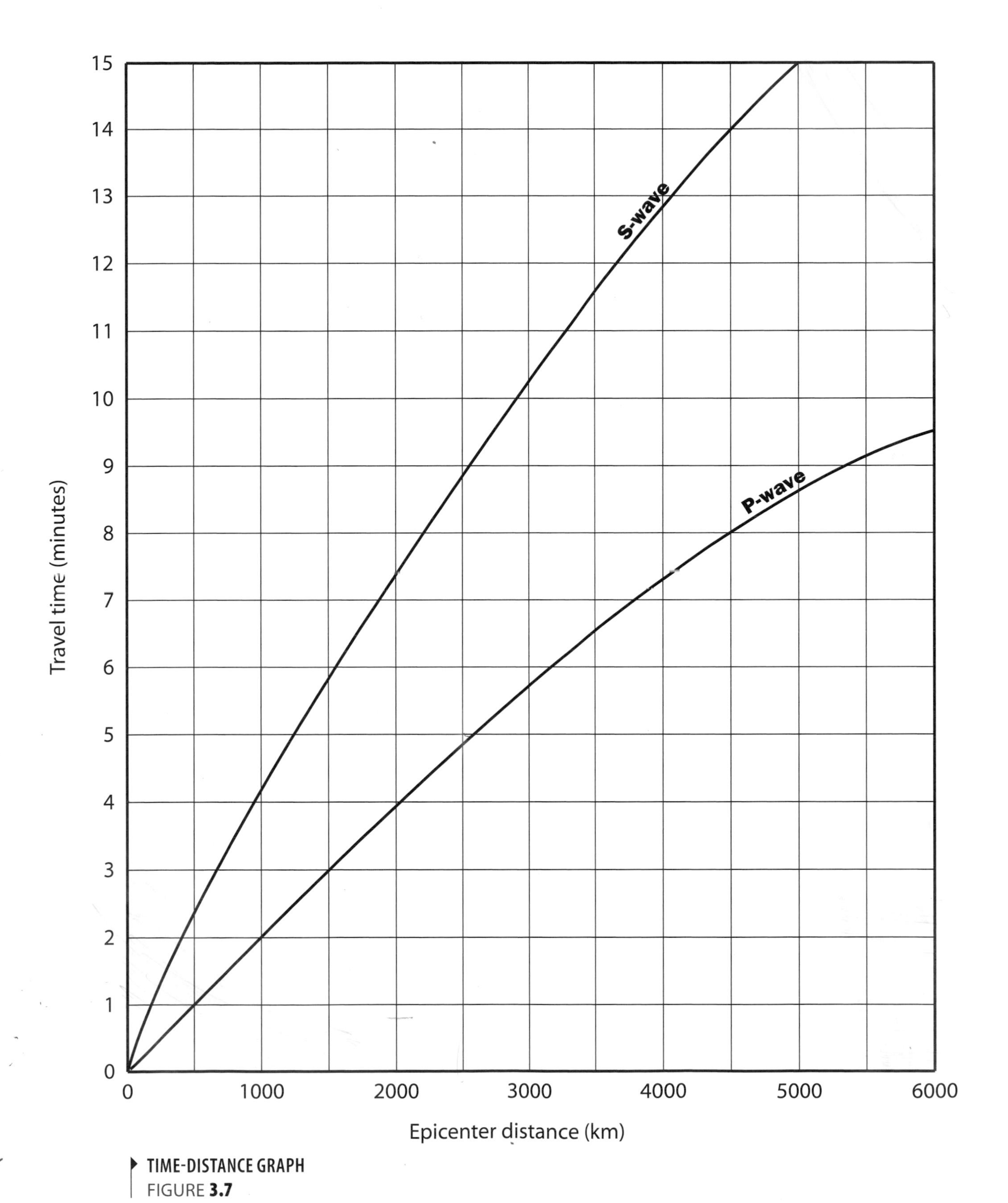

TIME-DISTANCE GRAPH
FIGURE **3.7**

Inquiry 3.3 *continued*

5 Review Procedure Steps 6 through 10 with your teacher.

6 You will use earthquake data that was recorded at three seismograph stations. Your job is to find out where the earthquake occurred, using the time-distance graph in Figure 3.7 and the seismograph data on Student Sheet 3.3.

7 Use the seismogram records in Figure 1 on Student Sheet 3.3 to complete Table 1 and answer these questions:

A. In what three cities was the earthquake recorded?

B. When did the P-waves first arrive at Station A (Sitka, Alaska)?

C. When did the S-waves arrive at Sitka?

D. How many minutes elapsed between the time the P-waves reached Sitka and the time the S-waves reached Sitka?

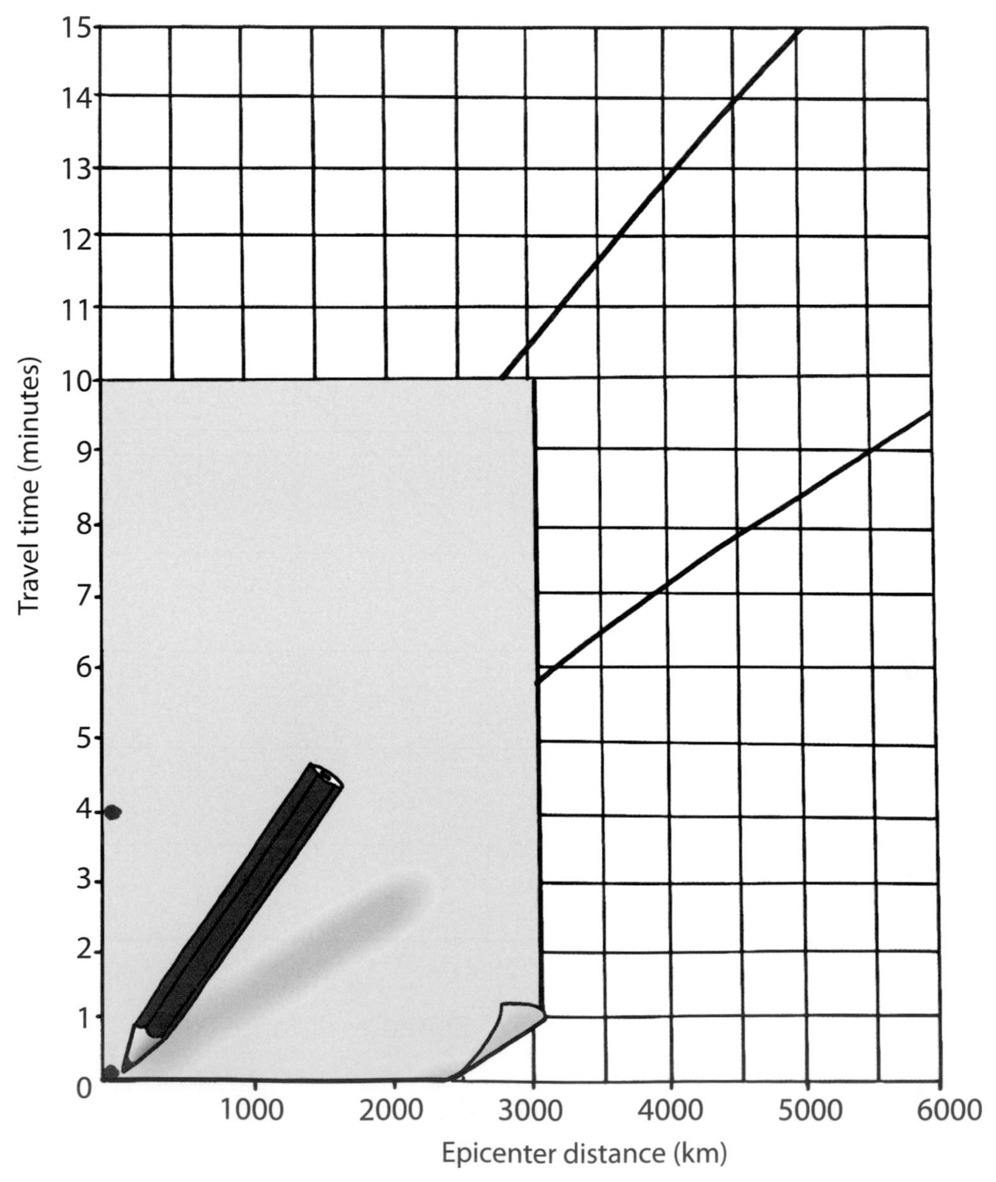

MARK TWO DOTS: ONE AT THE 0 MARK AND ONE AT THE 4-MINUTE MARK
FIGURE **3.8**

E. Determine the P- and S-wave arrival times for Stations B (Charlotte) and C (Honolulu). Compute the difference between these wave times. Complete column four of Table 1 on Student Sheet 3.3.

8 Use the S-wave minus P-wave data to find out how far the earthquake was from each station.

A. First, lay an index card along the vertical (time) axis of the graph in Figure 3.7.

B. Mark two dots on the edge of the index card for Station A (Sitka). The first dot should be at the 4-minute mark (the time for Sitka), and the second dot at the 0 point, as shown in Figure 3.8.

C. Slide the index card along the curved lines. Stop when the two dots match the curved lines, as shown in Figure 3.9.

D. Read and record the distance on the horizontal axis of your graph on Table 1 of Student Sheet 3.3.

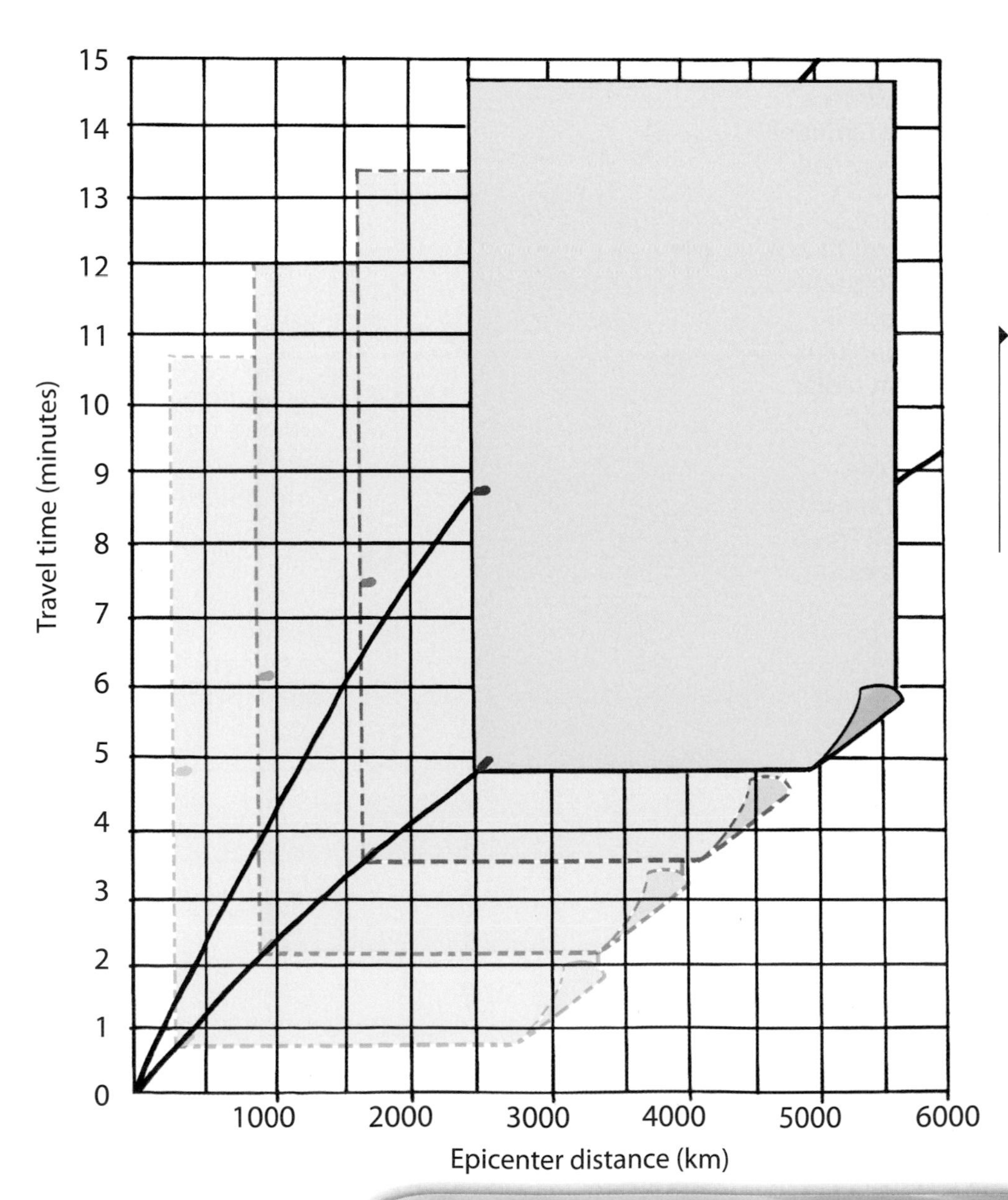

SLIDING THE INDEX CARD MARKED WITH FOUR MINUTES ACROSS THE CURVED LINES SHOWS THAT THE P- AND S-WAVE CURVES MATCH THE DOTS AT THE 2500-KM MARK.
FIGURE **3.9**

Inquiry 3.3 continued

9 Now repeat Step 8 for Station B (Charlotte) and Station C (Honolulu). Record your distances on Table 1 of Student Sheet 3.3.

10 Now you will use the map on Student Sheet 3.3 and your drawing compass to locate the epicenter for the earthquake.

A. Locate Sitka, AK, at Station A on the map. Remember the earthquake was 2500 km from Sitka (See Table 1 on Student Sheet 3.3).

B. Use the drawing compass to draw a circle centered at Sitka with a radius of 2500 km. The earthquake happened somewhere on this circle.

C. To find exactly where, you need to look at data from two other seismographs. Repeat steps A and B for Station B (Charlotte) and Station C (Honolulu), considering in each case what radius to use.

D. Locate the point where all three circles intersect (or come the closest to intersecting). Congratulations! This is the epicenter of the earthquake.

REFLECTING
ON WHAT YOU'VE DONE

1. With the class, review your completed Student Sheet 3.3 and results from this inquiry.

2. Draw conclusions about all three inquiries in this lesson. Answer these questions:

   A. What did you learn about how scientists record earthquakes?

   B. How does information on a seismogram tell scientists where earthquakes occur?

   C. How does knowing where earthquakes occur help people reduce the risks associated with future earthquakes?

3. Look ahead to Lesson 4. You will plot earthquake epicenters on a map to see whether there is a pattern in their locations.

Canines to the Rescue

In 1985, a violent earthquake shook Mexico City and destroyed a factory. The people working inside were trapped under the debris. After rescuers searched for 10 days, the Mexican government was ready to give up trying to find missing workers. As a last resort, relatives of the trapped victims asked rescue expert Carolyn Hebard to search the factory ruins with her dogs. Hebard flew in from her home in New Jersey with her specially trained German shepherds to help with the rescue operation. Hebard and her dogs worked their way into the wreckage and found two survivors.

A RESCUE DOG IS TRANSPORTED WITH OTHER RESCUE WORKERS ON A BLACKHAWK HELICOPTER TO SEARCH FOR RESIDENTS TRAPPED BY HURRICANE KATRINA.

PHOTO: FEMA/Jocelyn Augustino

READING SELECTION

EXTENDING YOUR KNOWLEDGE

HEBARD AND HER AMAZING DOGS

Hebard, a member of the National Association for Search and Rescue, is one of 1500 handlers of rescue dogs across the United States. She is one of the nation's respected experts on canine search and rescue.

Hebard first learned search-and-rescue techniques as a hobby, because she loved dogs and the outdoors. She joined a search-and-rescue club. With the club members, she trained her dogs, Pasha and Aranka, in the woods.

A RESCUE DOG ALERTS A WORKER THAT HE HAS POTENTIALLY FOUND THE SCENT OF A POSSIBLE HUMAN AMONG THE DAMAGE FROM HURRICANE KATRINA.

PHOTO: FEMA/Marvin Nauman

THE NOSE KNOWS

Dogs are good at finding people stranded under rubble because they have an excellent sense of smell. Instead of tracking a scent by sniffing the ground, rescue dogs are trained to sniff the air. The human body sheds tiny particles of dead skin that carry the human scent. These particles float on air currents where dogs' noses can easily detect them. By sniffing the air, dogs can search a much larger area than they could by just sniffing the ground.

Between search-and-rescue jobs, Hebard trains her dogs in the woods about once a week, just as athletes train between competitions. The dogs often practice by searching for one of Hebard's friends who has agreed to hide in the woods. As they sniff the air, the dogs catch a whiff of the person hiding. They know that they have a human scent and begin to zero in. Crouching low to the ground, they pace back and forth, smelling for the strongest scent.

THIS RESCUE DOG IS SEARCHING FOR PEOPLE TRAPPED BY HURRICANE KATRINA.

PHOTO: FEMA/Jocelyn Augustino

Following a scent is not always simple for dogs, says Hebard. The wind can confuse them. For instance, at an apartment building that collapsed after the 1995 earthquake in Kobe, Japan, three Swiss rescue dogs repeatedly sniffed and barked at the same two corners. Japanese rescue workers pried off a roof and moved other debris. After working for six hours, they found nothing. When they brought the dogs in again, the dogs indicated a different corner. The wind had been blowing from one direction in the morning and from a different direction in the afternoon.

THIS DOG IS SEARCHING FOR A VICTIM IN THE RED RIVER AFTER FLOODING IN NORTH DAKOTA IN 2009.

PHOTO: FEMA/Andrea Booher

A RESCUE WORKER AND RESCUE DOG SEARCH FOR SURVIVORS IN A PILE OF RUBBLE THAT RESULTED FROM A SERIES OF TORNADOES IN TENNESSEE IN FEBRUARY 2008.

PHOTO: FEMA/Jocelyn Augustino

HELPING OUT IN KOBE

Hebard and her dogs also helped out during the earthquake in Kobe. She and her dogs searched the ruins of the city's train station. "The railroad tracks actually looked like a roller coaster," she said. "It was incredible the way they had bent. We had to crawl inside the station." They climbed over piles of broken glass and crushed filing cabinets. It took four hours to crawl through a structure that would have taken 10 minutes to walk through before the earthquake.

When a search-and-rescue mission is over, Hebard and her dogs return to a quiet life in New Jersey. Sometimes it takes the dogs a little time to adjust to having other family members around.

"You do become very close to the dog, and definitely after you've been on an extended search," said Hebard. "The bond between you and that dog becomes closer and closer." ■

DISCUSSION QUESTIONS

1. Suppose you were asked to build an automatic "sniffer" that could do the same job as a rescue dog. How would you design it?
2. While dogs are useful in search-and-rescue efforts, the work remains costly and dangerous. How could the need for such rescues after earthquakes and other natural disasters be reduced?

LESSON 4

PLOTTING EARTHQUAKES

WORKERS LOOK FOR SURVIVORS IN A BUILDING DEVASTATED BY AN EARTHQUAKE IN MEXICO CITY, MEXICO, IN SEPTEMBER 1985.

PHOTO: U.S. Department of Defense

INTRODUCTION

On September 19, 1985, a strong earthquake occurred in Mexico City. The quake killed more than 9000 people and destroyed thousands of buildings. Two months later, a powerful volcanic eruption occurred about 3200 kilometers south of Mexico City. Do you think there could be a relationship between where earthquakes and volcanoes occur?

In this lesson and in Lesson 5, you will conduct inquiries that will help you begin to answer this question. In this lesson, you will plot on a world map a set of earthquakes that occurred in various parts of the world in the 1990s. You will then examine the map to determine whether you can see a pattern in the locations of the earthquake epicenters. Then, in Lesson 5, you will plot on the same map the locations of recent volcanic eruptions. Finally, you will analyze the locations of earthquakes and volcanoes. What do these powerful phenomena have in common? Let's find out.

OBJECTIVES FOR THIS LESSON

- Plot on a world map the locations, depths, and magnitudes of some of the earthquakes that occurred during the 1990s.
- Analyze the locations, depths, and magnitudes of earthquakes around the world.
- Locate three areas of intense earthquake activity on a map.
- Hypothesize about the reasons for patterns in the locations of earthquakes.

MATERIALS FOR LESSON 4

For you

1 copy of Inquiry Master 1.1: Plate Tectonics World Map

For your group

1 transparency copy of Inquiry Master 4.1: World Map
1 blue transparency marker
10 removable blue dots
1 laminated Plate Tectonics World Map from Lesson 1
1 paper towel

GETTING STARTED

1 Look at the brainstorming list your teacher posted from Lesson 1.

A. Do any of the statements on the list tell where earthquakes occur?

B. If not, where do you think most earthquakes occur? Your teacher will add your ideas to the list.

2 When asked, get the Plate Tectonics World Map that your group used in Lesson 1.

A. Where did your group record the most earthquakes?

B. Why did you think that earthquakes occur more frequently in these locations? Your teacher will add your ideas to the brainstorming list.

LAKE SAREZ IN TAJIKISTAN WAS FORMED WHEN AN EARTHQUAKE TRIGGERED A MASSIVE LANDSLIDE NEARLY 100 YEARS AGO. THE LAND DAMMED UP THE MURGHOB RIVER, FORMING THE LAKE. SCIENTISTS MONITOR THE AREA BECAUSE THEY WORRY THAT THE FREQUENT SEISMIC ACTIVITY MAY ONE DAY CAUSE THE BANKS TO SLUMP INTO THE LAKE, CREATING A HUGE WAVE THAT COULD OVERTOP THE DAM AND FLOOD AREAS DOWNSTREAM.

PHOTO: NASA Goddard Flight Center

INQUIRY 4.1

PLOTTING EARTHQUAKES TO IDENTIFY PATTERNS

PROCEDURE

1. Examine Table 4.1: Earthquake Data Set on pages 52–53. Listen as your teacher describes how you will use the lines of latitude and longitude to plot the earthquake epicenters on your group's world map and on a transparency.

2. Find the word "Magnitude" on Table 4.1. What do you think it means? Discuss your ideas with the class. Then read "Magnitude and Intensity" on pages 56–59. Give the magnitude of three different earthquakes. Of the three you chose, which earthquake released the most energy?

3. Find the word "Depth" on Table 4.1. What do you think it means? Give three examples of the depth of different earthquakes from Table 4.1.

4. Obtain one transparency copy of Inquiry Master 4.1: World Map, one blue transparency marker, and 10 blue dots for your group.

5. You and your group will be assigned 10 earthquakes from Table 4.1. Have two members of your group plot the epicenter of each of the 10 earthquakes on the transparency using the blue marker, while the other two members use the blue dots to plot those same earthquakes on the laminated Plate Tectonics World Map from Lesson 1. If you are working on the laminated Plate Tectonics World Map, write the Earthquake Number from the table on the blue dot. This will help you tell the difference between the earthquakes you plot in this lesson and those you plotted in Lesson 1.

Inquiry 4.1 continued

TABLE 4.1 EARTHQUAKE DATA SET

EARTHQUAKE NUMBER	DATE	LATITUDE	LONGITUDE	DEPTH (km)	MAGNITUDE ON RICHTER SCALE	LOCATION
1	1/25/09	43.3° N	80.8° E	19	5.3	Kazakhstan-Xinjiang border
2	10/7/09	13.0° S	166.2° E	35	7.7	Vanuatu Islands
3	12/19/09	23.7° N	121.7° E	45	6.4	Taiwan
4	12/19/09	23.8° N	121.7° E	45	6.4	East of Taiwan
5	2/26/10	25.9° N	128.4° E	22	7.0	Ryukyu Islands
6	2/27/10	35.9° S	72.7° W	35	8.8	Off coast of Chile
7	3/11/10	34.3° S	71.9° W	11	6.9	Chile
8	4/4/10	32.3° N	115.3° W	10	7.2	Northern Baja California
9	6/30/10	16.5° N	97.8° W	20	6.2	Acapulco coast, Mexico
10	7/7/10	33.4° N	116.5° W	14	5.4	Southern California, U.S.
11	8/5/10	43.6° N	110.4° W	5	4.8	Wyoming, U.S.
12	8/14/10	12.2° N	141.4° E	22	6.6	NW of Mariana Island
13	8/22/10	37.5° N	20.3° E	16	5.6	Ionian Sea
14	10/23/10	63.5° N	23.7° W	10	4.8	Iceland
15	11/4/10	12.8° N	44.9° W	10	5.6	Northern Mid-Atlantic Ridge
16	12/8/10	56.4° S	25.9° W	17	6.5	South Atlantic
17	1/10/11	23.1° N	143.2° E	73	5.7	Volcano Islands region
18	1/18/11	28.7° N	63.9° E	68	7.2	Pakistan
19	3/1/11	29.6° S	112.1° W	10	6.0	Easter Island region
20	3/6/11	56.4° S	27.0° W	84	6.5	South Sandwich Islands region

EARTHQUAKE NUMBER	DATE	LATITUDE	LONGITUDE	DEPTH (km)	MAGNITUDE ON RICHTER SCALE	LOCATION
21	3/22/11	33.1° S	16.0° W	11	6.1	Tristan de Cunha region
22	4/7/11	17.4° N	94.0° W	167	6.5	Southern Mexico
23	5/14/11	32.9° S	22.1° E	5	4.1	South Africa
24	5/15/11	0.5° N	25.6° W	10	6.0	Central Mid-Atlantic Ridge
25	7/26/11	2.7° S	76.6° W	124	5.2	Peru–Ecuador border
26	7/27/11	10.7° N	43.4° W	6	5.9	Northern Mid-Atlantic Ridge
27	8/1/11	34.7° N	138.5° E	16	6.2	Japan
28	8/13/11	14.5° N	94.7° W	28	5.6	Off coast of Southern Mexico
29	9/9/11	49.5° N	127.0° W	23	6.4	Vancouver Island region
30	9/24/11	7.6° S	74.5° W	145	6.8	Peru–Brazil border
31	10/13/11	43.4° N	127.2° W	10	5.3	Off coast of Oregon
32	10/19/11	38.0° N	31.4° W	11	4.9	North Atlantic Ocean
33	10/28/11	40.6° S	126.4° E	10	4.7	South of Australia
34	10/28/11	14.5° S	76.0° W	24	6.9	Coastal Peru
35	10/31/11	52.4° N	177.9° E	152	5.8	Aleutian Islands
36	11/18/11	37.6° S	179.3° E	26	6.0	Near North Island, New Zealand
37	11/22/11	15.4° S	65.1° W	557	6.6	Bolivia
38	11/25/11	63.4° N	150.5° W	150	3.0	Alaska, U.S.
39	11/29/11	1.7° S	15.4° W	10	5.9	East of Ascension Island
40	12/1/11	23.8° N	103.6° W	14	4.6	Mexico City, Mexico

Inquiry 4.1 continued

6 Share your results with the class by showing them the data points on both your transparency and your Plate Tectonics World Map.

7 Discuss patterns in the locations of these earthquakes. Answer the following questions, and support your answers with evidence from your map:

A. On (or near) which coast of North America do most earthquakes occur?

B. Which states in the United States are most earthquake-prone?

THE RING OF FIRE, OR CIRCUM-PACIFIC BELT, CONSISTS OF A CHAIN OF EARTHQUAKES AND VOLCANOES AROUND THE EDGES OF THE PACIFIC OCEAN. DEEP, NARROW DEPRESSIONS IN THE SEAFLOOR—CALLED DEEP-SEA TRENCHES—CIRCLE THE PACIFIC OCEAN ALONG THE RING OF FIRE.
FIGURE **4.1**

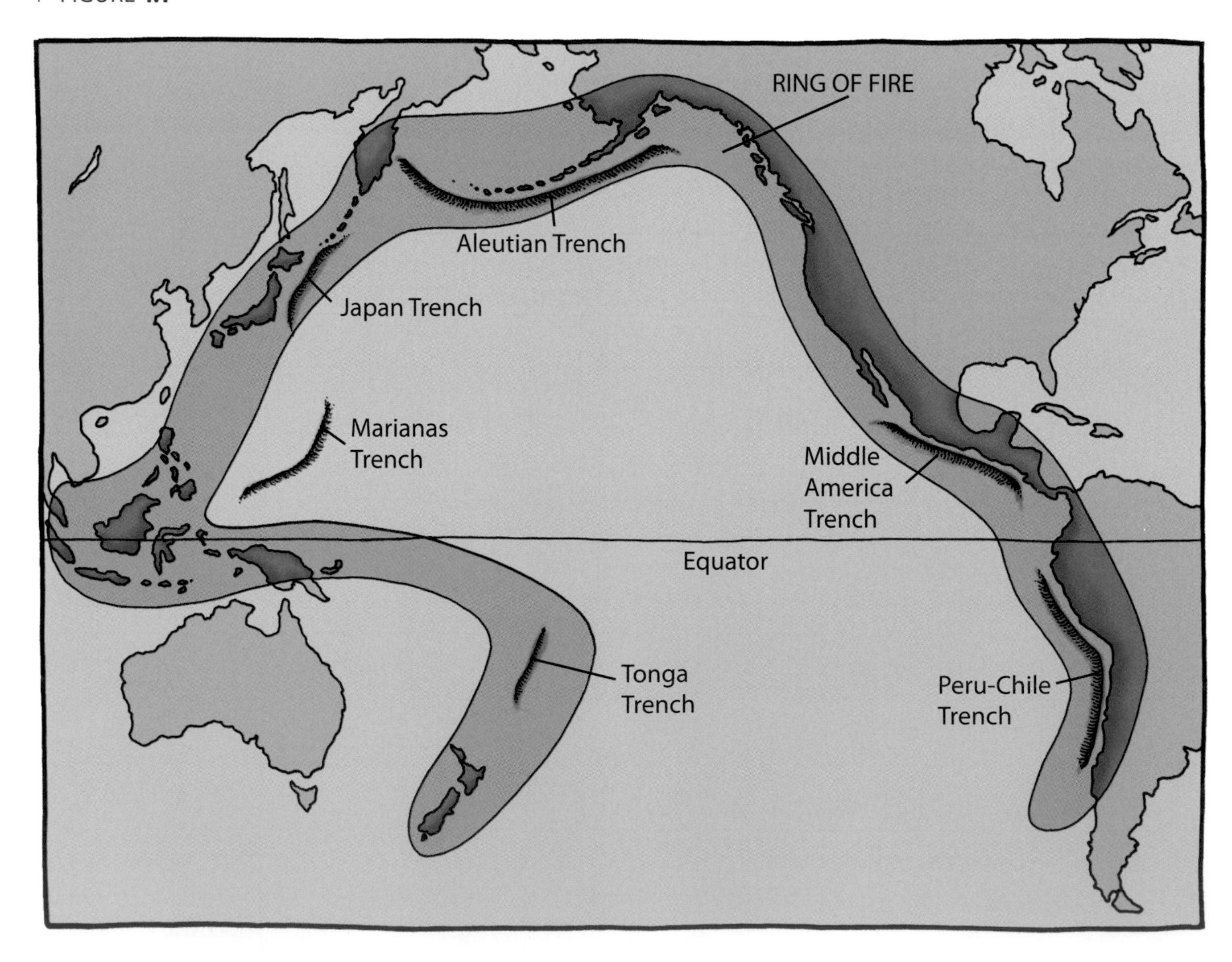

C. Are the earthquakes located within specific areas or scattered throughout the world?

D. If the earthquakes are in specific areas, how would you describe those areas?

8. Discuss with your class why it is important for scientists to identify patterns of earthquake activity.

9. Your teacher will show you Color Image 4.1: Seismic Activity Around the World. Which areas seem to have the most earthquakes? Share your ideas with the class. Your teacher will outline these areas of intense earthquake activity on the transparency.

10. Now look at Color Image 4.2. Seismic Belts, which your teacher will show you. On the transparency, find the Mid-Atlantic Ridge, the Mediterranean-Himalayan Belt, and the Circum-Pacific Belt (or Ring of Fire). (Figure 4.1 shows where the Ring of Fire is.) These are areas of intense earthquake activity. Label these areas on your group's world map.

REFLECTING ON WHAT YOU'VE DONE

1. Think about the patterns you outlined on your map. Answer these questions:

   A. Are there any earthquakes near mountain ranges? Continental coasts? Ocean basins? Trenches? Volcanic islands? Give specific examples for each one.

   B. Why do you think earthquakes are located in these specific areas? Think about what might be causing the earthquakes to occur.

   C. Look at the position of the earthquakes you plotted and the depths of the earthquakes. Are there any relationships that you can identify? If so, what do you think this might mean?

2. What causes patterns in earthquake activity? Your teacher will place Color Image 4.3: Plate Boundaries over Color Image 4.1, which shows seismic activity around the world. What do these two maps together tell you? The outer layer of the earth is broken into segments called plates. Earthquakes and volcanic activity are closely related to the movement of these plates.

3. Compare the locations of the earthquakes you plotted in Lesson 1 with those plotted in today's lesson. Do you want to revise any of your original ideas? Use your copy of Inquiry Master 1.1 to document your new ideas. Be sure to use a different color pen to show your new thinking.

4. Read "Using a Network for Seismic Monitoring" on pages 62–65. Looking at your Plate Tectonics World Map, discuss with your class where a seismic monitoring system might want to locate its sensors and stations.

5. Clean up. Use a paper towel to wipe all marks off the transparency.

6. Look ahead to Lessons 5–8. You will be learning more about movement along plate boundaries and why earthquakes occur.

READING SELECTION

EXTENDING YOUR KNOWLEDGE

MAGNITUDE & INTENSITY

Geologists measure an earthquake in two ways—by its magnitude and by its intensity. Each method provides these scientists and others with important data about the earthquake and its effects. Geologists can use the data to assess the risks in earthquake-prone regions and prepare for future earthquakes.

MAGNITUDE

"An earthquake with a magnitude of 6.8 on the Richter scale occurred today ..." How many times have you heard a number like this being reported in the news?

IN FEBRUARY, 2010, AN EARTHQUAKE IN CONCEPCIÓN, CHILE, SEVERELY DAMAGED THIS BUILDING. THE EARTHQUAKE REGISTERED 8.8 ON THE RICHTER MAGNITUDE SCALE.

PHOTO: U.S. Geological Survey/photo by Walter D. Mooney, Ph.D.

In 1935, Charles Richter, a seismologist at the California Institute of Technology, developed the Richter Magnitude Scale. The Richter scale measures the magnitude, or total amount of energy, released at the source of an earthquake. The number that you normally hear on the news when an earthquake occurs is its magnitude. Richter scale ratings enable people to compare the strength of different earthquakes around the world.

The magnitude of an earthquake is determined by measuring the amplitude, or "swing," of the largest seismic wave on a seismogram. The Richter scale, shown in the table below, is open-ended; it has no maximum magnitude. As of the year 2010, the largest magnitude recorded on the Richter scale was 9.5. That earthquake occurred in Chile in 1960. The largest earthquake in the U.S. occurred in Alaska in 1964. It registered 9.2 on the Richter scale.

Each increase in a magnitude number on the Richter scale represents a tenfold increase in the amplitude seen on the seismogram. This means that a magnitude-6 earthquake has an amplitude 10 times greater than a magnitude-5 earthquake and 100 times greater than a magnitude-4 earthquake. This greater amplitude translates into longer and higher energy shaking of the ground. For example, an earthquake with a magnitude around 5.0 might only shake the ground for 30 seconds or so, while the 9.2 Alaska earthquake shook the ground for over nine minutes. And for every increase of magnitude of 1.0, there is an increase of 32 times the amount of energy released.

RICHTER MAGNITUDE SCALE

DESCRIPTOR	MAGNITUDE	AVERAGE NUMBER EACH YEAR, WORLDWIDE
GREAT	8 and higher	1
MAJOR	7–7.9	18
STRONG	6–6.9	120
MODERATE	5–5.9	800
LIGHT	4–4.9	6200 (estimated)
MINOR	3–3.9	49,000 (estimated)
VERY MINOR	Less than 3.0	Magnitude 2–3: about 1000 per day Magnitude 1–2: about 8000 per day

SOURCE: National Earthquake Information Center, U.S. Geological Survey, Denver, CO

READING SELECTION

EXTENDING YOUR KNOWLEDGE

INTENSITY

Scientists use the word "intensity" to describe the kind of damage done by an earthquake, as well as people's reaction to the damage. In other words, intensity is a measure of the earthquake's effect on people, structures, and the natural environment.

Many factors affect intensity. These include the distance an area is from the epicenter, the depth of the earthquake, the population density of the area affected by the earthquake, the local geology of the area, the type of building construction in the area, and the duration of the shaking. Magnitude also affects intensity, since an earthquake of a higher magnitude has a higher intensity than an earthquake of lower magnitude. But, an earthquake in a densely populated area that results in many deaths and considerable damage (high intensity) may

MODIFIED MERCALLI INTENSITY SCALE

INTENSITY SCALE	DAMAGE AND FELT OBSERVATIONS
I	Not felt, except by a very few people under special circumstances.
II	Felt only by a few people at rest, especially on upper floors of buildings.
III	Felt only indoors, but many people did not recognize it as an earthquake. Stationary cars rocked slightly.
IV	Felt indoors by many, outdoors by few. At night some people were awakened by a sensation like a heavy truck hitting a building. Standing cars rocked noticeably.
V	Felt by nearly everyone. Many were awakened. Some dishes and windows were broken. Trees and other tall objects swayed.
VI	Felt by all. Many were frightened and ran outdoors. Heavy furniture moved. Plaster on walls and chimneys was damaged.
VII	Everyone ran outdoors. Slight to moderate damage to well-built structures. Considerable damage to poorly built structures. Some chimneys broken. Noticed by people driving cars.
VIII	Damage slight in well-designed structures, great in poorly built structures. Fallen chimneys, monuments, and walls. Heavy furniture was overturned. Sand and mud were ejected from the ground in small amounts.
IX	Damage was considerable in well-designed structures. Buildings shifted off foundations. Ground noticeably cracked. Underground pipes broken.
X	Well-built wooden structures destroyed. Ground badly cracked. Railroad tracks bent. Landslides considerable. Water splashed over riverbanks.
XI	Few, if any, masonry structures remained standing. Bridges were destroyed. Broad cracks formed in ground. Underground pipes completely out of service.
XII	Total damage. Waves seen on ground surfaces. Objects thrown upward into the air.

A 2008 EARTHQUAKE IN SICHUAN, CHINA, CAUSED A SECTION OF THIS BUILDING TO COLLAPSE.

PHOTO: U.S. Geological Survey/photo by Sarah C. Behan

have the same magnitude as an earthquake in a remote area that does no more than frighten the wildlife (low intensity).

The most common earthquake intensity scale used in the United States is shown in the table titled "Modified Mercalli Intensity Scale." This scale has intensity values ranging from I to XII. (Can you guess why it uses Roman numerals?) ■

DISCUSSION QUESTIONS

1. Look at the photo of earthquake damage in Sichuan, China. Where would you put this earthquake on the intensity scale?
2. On average, there are about 8000 earthquakes of magnitude 1–2 per day, and only one earthquake of magnitude 8 or higher each year. Why are there so many minor earthquakes and so few major ones?

READING SELECTION

EXTENDING YOUR KNOWLEDGE

USING HISTORICAL EARTHQUAKE INTENSITY TO ESTIMATE FUTURE RISK

Since the mid-19th century, well before the Richter Magnitude Scale was developed, scientists have been using intensity as an approximation of the strength of an earthquake. The U.S. Geological Survey (USGS) collects data about historical earthquakes based on "felt observations" of citizens. People who think they have felt the impacts of an earthquake can go to the USGS website and fill out a questionnaire about where they were and what they experienced. Compiling this data helps the USGS determine the intensity of earthquakes.

If records existed of the "felt observations" for all the earthquakes that have occurred on the earth, the maximum intensity of all earthquakes at a particular site would be a good estimate of earthquake risk for that area. But scientists only have a small amount of data about the earthquakes that have occurred throughout history. Therefore, they make intelligent guesses about where and how often earthquakes occur, how large they will be, and how much shaking will occur. This information is then plotted on a map. People moving into an earthquake-prone city might use the map to get some idea of the risk of an earthquake occurring in that area. Earthquake maps are also used to determine building codes and insurance rates. ■

THESE TWO PHOTOS SHOW DAMAGE FROM THE SAN FRANCISCO EARTHQUAKE OF 1906, WHICH REGISTERED 7.6 ON THE RICHTER MAGNITUDE SCALE. WHAT DO YOU THINK THE INTENSITY OF THIS EARTHQUAKE MIGHT HAVE BEEN?

PHOTO (right): Library of Congress, Prints & Photographs Division, LC-USZ62-17359
PHOTO (below): Library of Congress, Prints & Photographs Division, LC-USZ62-64748

DISCUSSION QUESTIONS

1. "Felt observations" have proven useful for understanding historical earthquakes. If you were guiding people today to keep a record of their felt observations, what information would you ask them to record, and in what form?

2. Besides a map showing felt observations, what other data could people use to get an idea of earthquake risk in a particular area?

READING SELECTION

EXTENDING YOUR KNOWLEDGE

network Search

Using a Network for Seismic Monitoring

Do you use social networks through your computer, phone, or other device to keep in touch with your friends? Maybe you share information such as photos, or news about people you know. Scientists around the world who study earthquakes and volcanoes also use digital networks to keep in constant contact with each other, sharing scientific news, data, and ideas. The information they are sharing is about one of their favorite topics, seismic activity (frequency, intensity, and magnitude of earthquakes). Your conversations and the scientists' communications both rely on telecommunications networks, including satellites, sensor networks, and the Internet.

The Global Seismographic Network is a permanent set of more than 150 sensors that are placed around the world to monitor seismic activity. They are connected by a telecommunications network that allows data to flow from the sensors to scientists who can monitor it. The network was conceived in 1984 by a group of organizations, led by the U.S. Geological Survey (USGS), which were trying to coordinate the monitoring of seismic activity. While several organizations around the globe had been monitoring seismic activity, they had no easy way to share their information to produce timely, global analyses of Earth's crust in motion. With the Global Seismographic Network, not only did they succeed in connecting themselves in a network, they were able to get funding to set up more seismic monitoring stations. Today, sensors are in locations as diverse as a dairy farm in Singapore, a snowy peninsula in Antarctica, and a remote island nation in the South Pacific called Tuvalu.

You may belong to some sort of club. As a member, you probably take part in regular club activities, like meetings or practices, and you agree to follow the club's rules or guidelines. Members of the Global Seismographic Network must agree to collect regular, standardized data on seismic activity. Each of the more than 100 members—universities, government agencies, and institutes representing about sixty different countries—maintains one or more sophisticated sensors. In exchange for keeping their sensors in good, working order, the members get access to worldwide data on seismic activity. Just as your club president should make sure the activities of the club are running smoothly, the USGS helps many of the member countries maintain their sensors.

Collectively, the sensors of the Global Seismographic Network provide nearly uniform coverage of seismic events on the earth. Each sensor measures and records vibrations, and the sensors' clocks are set to be highly accurate, using global positioning satellites (GPS). It is important that scientists know exactly when seismic activity is taking place, so that they can see and study all the simultaneous activity around the world. The data collected by each sensor is sent to earthquake information

WHY DO YOU THINK THE TINY ISLAND COUNTRY OF TUVALU WOULD AGREE TO MAINTAIN A SEISMIC SENSOR?

PHOTO: Stefan Lins/creativecommons.org

and warning centers, using satellites and the Internet. Some of the seismic data is sent in real time, which means people receive it right away, while the events are happening. This can be extremely useful. Knowing the epicenter and magnitude of an earthquake immediately can help warning centers mobilize a rapid emergency response. Someday, as our technology improves, all of the monitoring stations' data will be available in real time.

Even data that is not sent in real time can be used by researchers to better understand seismic events. Thanks to the Global Seismographic Network, within an hour of an earthquake anywhere on the globe, scientists know its location, depth, and magnitude. Using the data collected, scientists have been able to learn more about the earth's interior structure and what happens during an earthquake. They are looking at things such as the deep structure of spreading ridges. The more we can understand about the workings of the inner earth, the better chance we have of learning how to anticipate and prepare for earthquakes.

Data from the network also helps us understand the aftermath of earthquakes.

READING SELECTION

EXTENDING YOUR KNOWLEDGE

SOLAR PANELS ARE USED TO POWER THIS SEISMIC MONITORING STATION IN THE BRAZILIAN AMAZON.

PHOTO: Earthquake Hazards Program/U.S. Geological Survey

SENSORS WILL RECORD ANY VIBRATIONS, INCLUDING THOSE NOT CAUSED BY EARTHQUAKES.

PHOTO: Kyle Nishioka, kylenishioka.com/creativecommons.org

High-frequency vibrations occur during an earthquake, but we now know they are followed by low-frequency vibrations that shake the earth for days or weeks. The giant Sumatra earthquake, which lasted about seven minutes, set the entire planet to vibrating within 21 minutes.

The Global Seismographic Network is also used for the military purpose of detecting underground nuclear explosions. In fact, the U.S. Department of Defense was one of the founding partners of the network, and some of the money to maintain it comes from the U.S. Air Force. As early as the 1960s, scientists realized that nuclear explosions generated seismic waves: waves of energy that travel through the earth. The seismic waves produced by a nuclear explosion are large enough to be detected around the world. This means that countries can monitor each other to see who is testing nuclear weapons, sometimes in violation of international treaties. The U.S. Department of Defense operates a specialized center to scan data from the network for evidence of nuclear tests. It has documented a number of nuclear explosions, including the series of nuclear tests conducted by India and Pakistan in 1998.

You may wonder how scientists can tell the difference between a seismic wave generated by an earthquake and one generated by a nuclear weapons explosion. It turns out that while earthquakes produce strong waves along the earth's surface, the waves from nuclear explosions tend to be confined to deeper layers. Determining the kinds of waves produced by a seismic event allows for identification of its cause.

So, the Global Seismographic Network really serves three functions: earthquake monitoring, earthquake research, and nuclear test monitoring. The network does all three jobs best if the sensors are numerous, well-placed, reliable, and linked through strong communication networks. Both scientists and military strategists are working to improve the network and make it fully real-time, using state-of-the-art equipment. If we're lucky, the Global Seismographic Network's continuing work will mean that students in the future understand our shifting, elastic planet even better than top seismologists can today. ■

DISCUSSION QUESTIONS

1. Think about how your class might work on a large project: what are the advantages and disadvantages of working independently in small groups versus linking all groups' efforts together into a classroom effort?
2. What does this tell you about the advantages and disadvantages of linking individual monitoring networks into a global monitoring network?

RING A BELL AND THEN STOP ABRUPTLY, AND YOU'LL UNDERSTAND WHY THE CONTINUED VIBRATIONS AFTER AN EARTHQUAKE ARE CALLED "RINGING" OF THE EARTH.

PHOTO: Steven Brener/creativecommons.org

LESSON 5

USING EARTHQUAKES TO STUDY THE EARTH'S INTERIOR

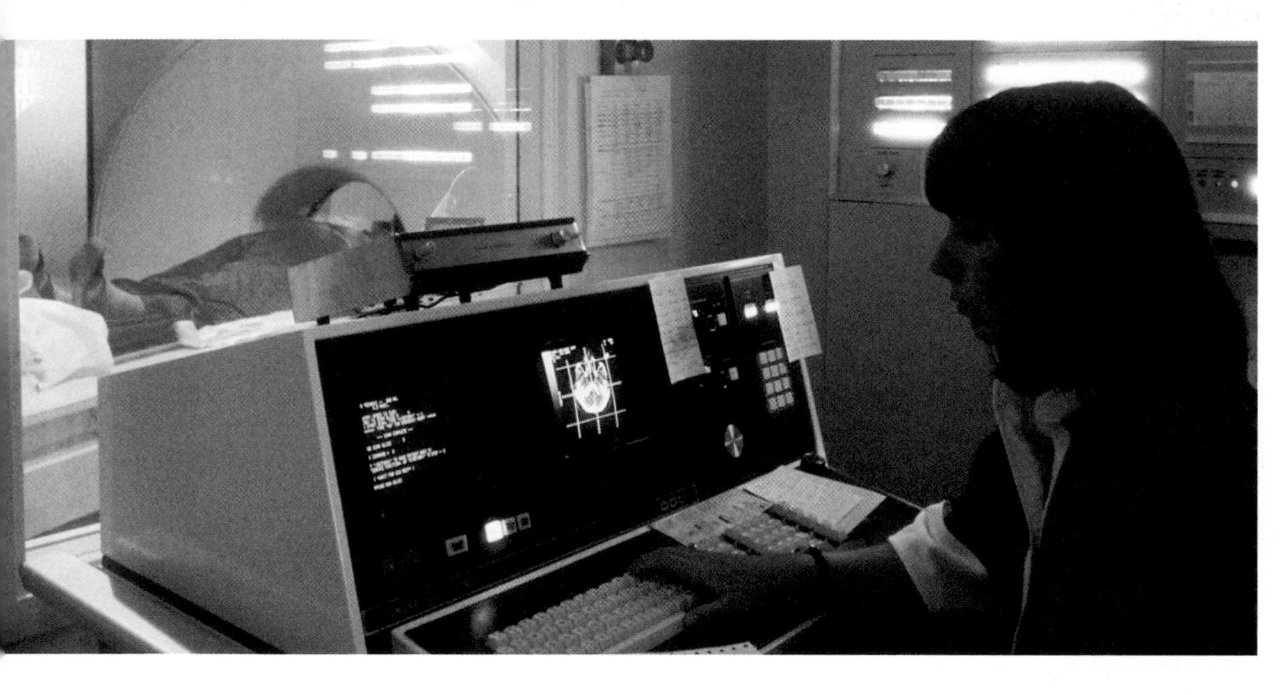

THIS TECHNICIAN IS EXAMINING IMAGES OF A PATIENT'S HEAD DURING A CT SCAN.

PHOTO: National Cancer Institute/Linda Bartlett

INTRODUCTION

Doctors can study the inside of the human body using a technique called computed tomography (CT). The CT scanner passes X-rays through the patient from different directions and creates three-dimensional images of the interior of the human body. An examination of these images helps the physicians diagnose diseases, disorders, and other health-related occurrences.

In a similar way, scientists use earthquake waves to learn more about the inside of the earth. How can earthquake waves help scientists better understand the earth and its interior? In this lesson, you will investigate this question and view computer images of the earth's layers. You will also read about each layer of the earth's interior.

OBJECTIVES FOR THIS LESSON

- Examine the interior structure of some common objects.
- Discuss how scientists study the structure of the earth's interior.
- Recognize that an understanding of the motion of earthquake waves can help scientists formulate hypotheses about the earth's interior.
- Use computer images to identify and describe the layers of the earth.
- Plot the locations of volcanoes and compare these locations with those of earthquakes.

MATERIALS FOR LESSON 5

For you

	Items with interior structure and/or a diagram (Homework from Lesson 4)
1	copy of Student Sheet 5.1: Plotting Volcanic Activity

GETTING STARTED

1. Share your homework with the class. Consider the following questions:

   A. How did you discover the internal structure of the object you chose?

   B. How does the structure of this object compare with that of objects other students brought to class?

   C. How did you use diagrams and labels to show the structure of the object you chose?

2. Look closely at the items your teacher shows you. With your group, discuss each object. These questions will guide you:

   A. What do you think each item looks like inside?

   B. How could you find out about the interior structure of each item?

3. Watch as your teacher cuts open each item. Compare the objects. How are they alike? How are they different?

4. Share what you already know and what you want to know about the earth's interior structure. Your teacher will record your ideas.

5. Discuss these questions with the class:

   A. How do you think the structure of these items resembles the structure of the earth's interior?

   B. How do you think scientists study and learn about the earth's interior?

SHORT OF CUTTING IT OPEN, HOW COULD YOU DETERMINE THE INTERNAL STRUCTURE OF AN APPLE?

PHOTO: Scott Bauer, Agricultural Research Service/U.S. Department of Agriculture

EXAMINING THE EARTH'S INTERIOR

PROCEDURE

1. Think back to Lessons 2 and 3. You used a steel spring to simulate earthquake waves and then used the different speeds of these waves to locate the epicenter of an earthquake. How do you think scientists use earthquake waves to tell whether the interior of the earth is solid, liquid, or gas?

2. Watch as your teacher creates waves in a tray of water. Then predict what might happen to the waves when a can is placed in the center of the tray. Discuss your prediction with the class. Next, watch as your teacher tests your prediction. Discuss your observations with the class.

3. Look at the diagram of the earth's interior on page 70. What can the motion of earthquake waves tell about the earth's interior?

4. Your teacher will use a CD-ROM to show you pictures of the earth's interior. Examine the images carefully. What do they tell you about the earth's interior?

REFLECTING ON WHAT YOU'VE DONE

1. As a class, answer the review questions from the CD-ROM.

2. Read "The Earth's Interior" on pages 70–71.

3. Answer these questions in your science notebook:

   A. Why are the images on the CD-ROM and in the reading selection drawings rather than photographs?

   B. Draw and describe each layer of the earth.

4. Read "Using Waves to Explore the Earth's Interior" on pages 72–73. How do earthquake waves help scientists learn more about the earth's interior? Use words and drawings to explain your ideas.

5. For homework, you will complete Student Sheet 5.1: Plotting Volcanic Activity to find out how the locations of volcanoes and earthquakes are alike.

6. Look ahead to Lessons 6, 7, and 8, in which you will investigate the causes of earthquakes and volcanoes as they relate to plate movement.

READING SELECTION

EXTENDING YOUR KNOWLEDGE

The Earth's INTERIOR

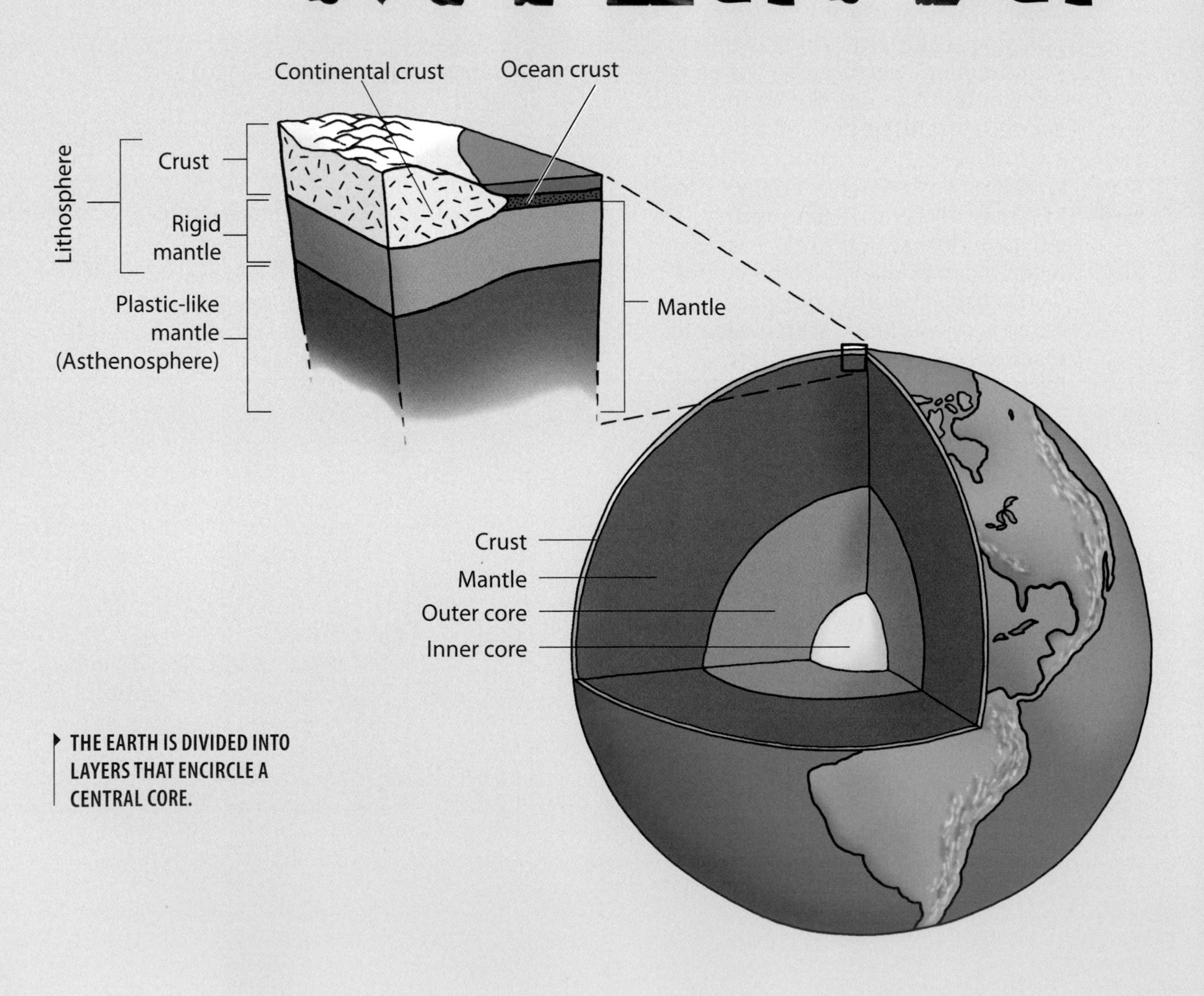

THE EARTH IS DIVIDED INTO LAYERS THAT ENCIRCLE A CENTRAL CORE.

Ever since its formation—some 4.5 billion years ago—the earth has been losing heat. The deeper one goes inside the earth, the greater the temperature becomes. The pressure rises, too. The earth's outer layer, or crust, is the coolest and least dense of all the layers inside the earth. (You might compare the earth with a loaf of bread that is cooling on a shelf. The crust cools first; the soft inner part of the loaf remains warm much longer.)

There are two kinds of crust: oceanic and continental. The oceanic crust lies beneath the ocean. It is approximately 5 to 10 kilometers (3 to 6 miles) thick. The continental crust contains mostly land. It ranges from 15 to 70 kilometers (9 to 43 miles) thick and is thickest under high mountain areas. Both types of crust are made up of rock.

Directly under the crust is the mantle. Like the crust, the mantle is composed of rock; however, the rock in the mantle is much denser than that in the crust. The mantle is about 2900 kilometers (1800 miles) thick, and it makes up about 83 percent of the earth's interior. The top layer of the mantle is rigid. It is cooler than the lower part of the mantle. Geologists call this rigid part of the mantle, together with the crust, the lithosphere. The lithosphere is broken into pieces, called "plates." (To visualize these plates, think about how an egg looks when its shell is cracked.)

The plates of the lithosphere "float" on the part of the mantle directly below it. This part of the mantle is called the asthenosphere. The consistency of the asthenosphere is like taffy. The asthenosphere is hot, and, like warm taffy, it can flow. The movement of the plates of the lithosphere on top of the slowly moving asthenosphere accounts for the formation of many mountains and volcanoes, as well as for earthquakes.

Beneath the mantle is the earth's innermost layer, the core. (Think of the center of an apple, which is also called the core.) The earth's core is divided into two parts: a liquid outer core, made of iron, and a solid inner core, made of iron and nickel. ■

DISCUSSION QUESTIONS

1. Why is the crust the coolest layer of the earth?
2. How is rock in the earth's mantle different from rock in the earth's crust?

READING SELECTION

EXTENDING YOUR KNOWLEDGE

USING WAVES TO EXPLORE THE EARTH'S INTERIOR

The deepest that scientists have drilled into the earth is 12 kilometers (7.5 miles). That's less than 0.2 percent of the distance from the surface of the earth to its center! So how do scientists know so much about the layers of earth's interior? How do they know, for example, that the lithosphere is rigid? Or that the asthenosphere is soft, like taffy? Or that the outer core is liquid?

To understand how scientists study the earth's interior, think about how they study the deepest parts of the ocean floor, which, like the depths of the earth, have never been explored directly by humans. Scientists study the ocean floor and the inner earth using waves. To study the ocean, they analyze sound waves, using a technique called sonar. To study the inside of the earth, they analyze earthquake, or seismic, waves.

SONAR WAVES

"Sonar" stands for Sound Navigation and Ranging. A sonar system consists of a transmitter and a receiver, just like a walkie-talkie, a phone, or any other two-way communication device. The sonar transmitter sends waves from a ship to the ocean floor. The waves bounce off the ocean floor, as shown in the illustration. A receiver detects the reflected waves. Oceanographers measure the time it takes for the sound waves to complete a round trip. Because they know how far sound can travel in a certain amount of time, the scientists can then determine the depth of a specific area of the ocean. They can also combine information from many sound waves to create a profile that shows the shape of specific areas of the ocean floor.

SOUND WAVES EMANATING FROM A SONAR SYSTEM ON A SHIP'S HULL COLLECT SONAR DATA IN A FAN-SHAPED AREA ON THE SEAFLOOR.

PHOTO: National Ocean Service/NOAA

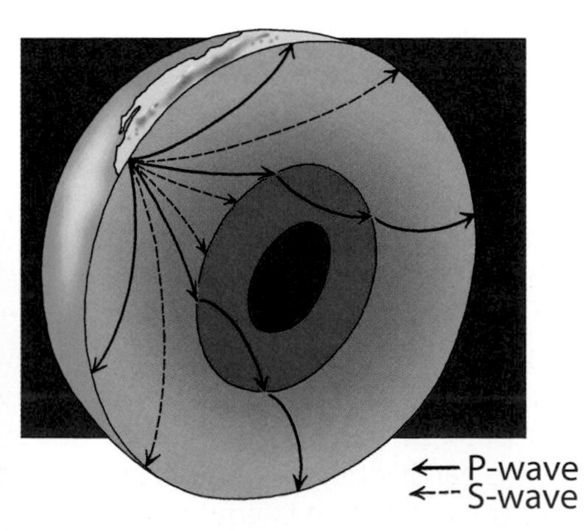

THE PATHS OF P- AND S-WAVES PROVIDE SCIENTISTS WITH INFORMATION ABOUT THE EARTH'S INTERIOR STRUCTURE. FOR EXAMPLE, S-WAVES, UNLIKE P-WAVES, DO NOT TRAVEL THROUGH LIQUIDS.

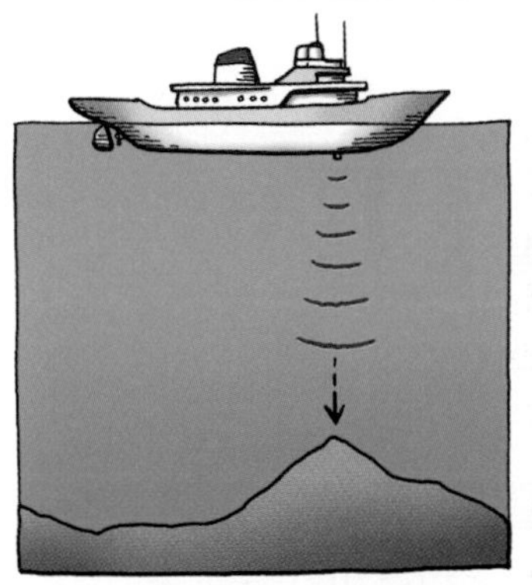

Sound pulse transmitted from ship

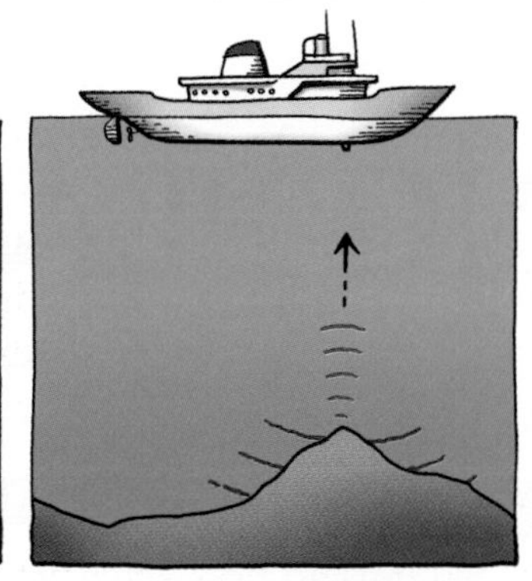

Sound wave hits and is reflected back from bottom

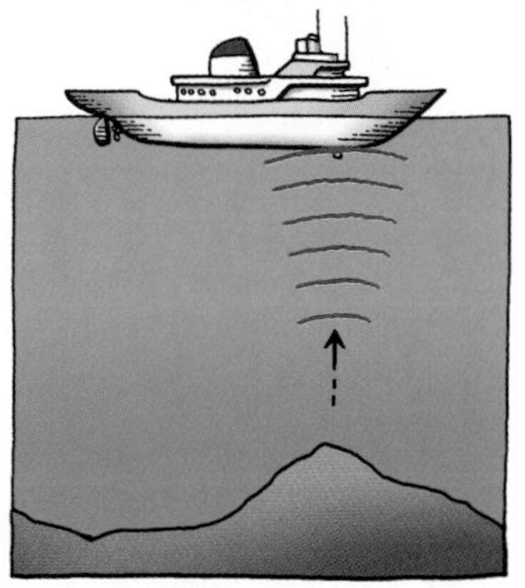

Reflected sound wave received by ship

SOUND WAVES BOUNCE OFF THE OCEAN FLOOR AND RETURN TO THE RECEIVER ABOVE.

OCEAN SCIENTISTS ABOARD A U.S. COAST GUARD SHIP LOWER SONAR EQUIPMENT INTO THE WATER FOR THEIR MAPPING OF THE SEAFLOOR NORTH OF ALASKA. THEY HOPE TO BETTER UNDERSTAND CURRENTS AND CLIMATE THROUGH THEIR RESEARCH.

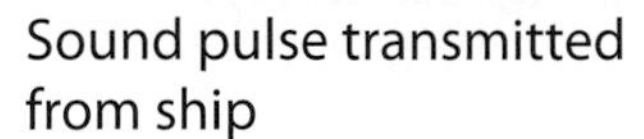

PHOTO: National Ocean Service/NOAA

EARTHQUAKE WAVES

Seismologists use earthquake waves to map the structure of the interior of the earth in much the same way oceanographers use sonar to map the ocean floor. As the earthquake waves move through the different layers of the earth, they change speed and direction. Sometimes they even stop. In other words, earthquake waves behave differently, depending on what substance they are traveling through. Because scientists know the average speed of P- and S-waves and also know how the waves travel, they can make educated guesses about the substances that make up the earth's interior. For example, they know that the outer core is liquid, because S-waves, which cannot travel through liquids, do not travel through the core.

The more scientists learn about sound waves and seismic waves, the more they may discover about the earth's most hidden area—its interior. ■

DISCUSSION QUESTIONS

1. Why do waves move at different speeds through different layers of the earth?
2. Why might S-waves travel more slowly than P-waves? Use Internet or library resources to see if your hypothesis is correct.

LESSON 6

INVESTIGATING PLATE MOVEMENT

A GLOBE IS A MODEL OF THE EARTH.

PHOTO: Anne Williams/©NSRC

INTRODUCTION

What does the word "model" mean to you? A model is often a smaller version of an object, such as a miniature airplane. In earth science, the shell of an egg can be used as a model of the earth's crust, which is too large and too complex to study firsthand in the classroom. In this lesson, you will use models to study how the earth's plates move and how the forces created by this movement result in earthquakes.

You will use two different styles of models in this lesson. One is quite simple, and the other is more complex. Using these models, you will investigate how the earth's lithospheric plates collide, separate, and slide past one another. You will use a relief globe to look for evidence of these interactions in landforms on the earth.

OBJECTIVES FOR THIS LESSON

- Describe what a model is and distinguish models from real objects or events.
- Use models to simulate the movement of lithospheric plates as they collide, separate, and slide past one another.
- Use a globe and a map to find evidence of plate movement and to identify landforms that result from plate movement.

MATERIALS FOR LESSON 6

For you

1 completed copy of Student Sheet 5.1: Plotting Volcanic Activity

For your group

2 thin foam pads
2 thick foam pads
2 desks, thick books, or other thick flat surfaces
1 laminated Plate Tectonics World Map (from Lesson 4)
1 relief globe
1 metric ruler

GETTING STARTED

1 Look at the Earth's Fractured Surface wall map with your class.

A. How is this map different from others you have seen?

B. How is it the same?

2 Review your homework (Student Sheet 5.1) from Lesson 5 with your teacher. You may be asked to plot a volcano data point on the wall map.

A. Describe any patterns you notice in the location of volcanoes and earthquakes.

3 Look at the plates shown on the map.

A. Describe any relationship you see between the locations of earthquakes and volcanoes on the earth and the boundaries, or outer edges, of plates.

B. What ideas do you have to explain this relationship?

4 View the CD-ROM *The Theory of Plate Tectonics.*

A. How do plates move?

B. How does plate movement cause earthquakes, volcanoes, mountains, and trenches?

C. Discuss your observations with your class.

5 Discuss what you know about models with your class.

6 Look at the models you will use in this lesson.

A. Which kind of plate boundary do you think you will model with each one?

B. Discuss with your class why and how models are used in the science classroom.

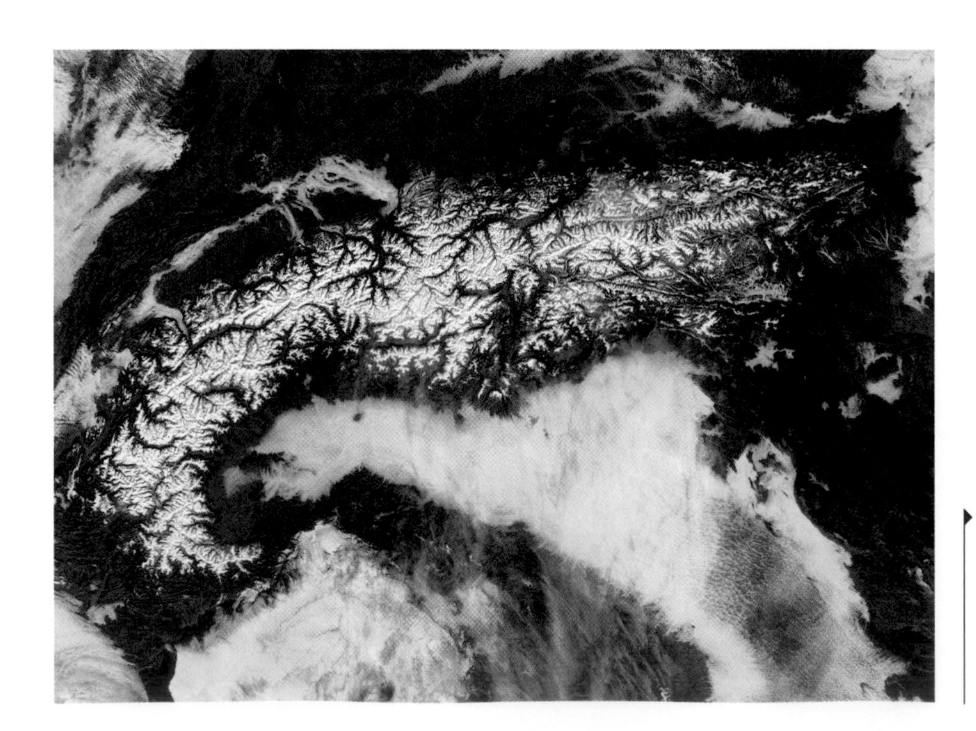

MOVING PLATES CAUSED THE ALPS MOUNTAIN RANGE THAT RUNS THROUGH EUROPE TO FORM. CAN YOU FIGURE OUT WHICH PLATES?

PHOTO: NASA image courtesy Jeff Schmaltz, MODIS Rapid Response Team at NASA GSFC

INQUIRY 6.1

USING SIMPLE MODELS OF PLATE MOVEMENT

PROCEDURE

1. In your science notebook, set up a T-chart with "Properties" above the left column and "Comparison of thin and thick pads" above the right column. With your group, examine the thin and thick pads. Describe their properties. Consider density, appearance, thickness, weight, and size. Record your group's ideas.

2. Think about and then discuss your predictions of how you think each pad would respond if you did the following things:

   A. Gently pulled on the pad from opposite ends

   B. Pushed on the pad at opposite ends toward the center

   C. Slid two pads past one another

   D. Collided two pads by pushing them together

Inquiry 6.1 continued

3 For each of the following procedures, list the action you perform on each pad or set of pads, then draw the results. Sketch the pad or pads you use. Remember to use labels and arrows to communicate the action. Test your predictions. The questions listed below are to help you make better observations.

A. Pull on one pad. What happens to its appearance? What happens to its volume and size? Hold one foam pad by its opposite ends, as shown in Figure 6.1. Compress the pad by pushing your hands toward each other. Repeat this process with the other type of pad.

B. Now try stacking the pads on top of each other. Compress them. What do you observe?

C. Place two pads of the same thickness side by side on the desk or table. Slide them past one another, as shown in Figure 6.2.

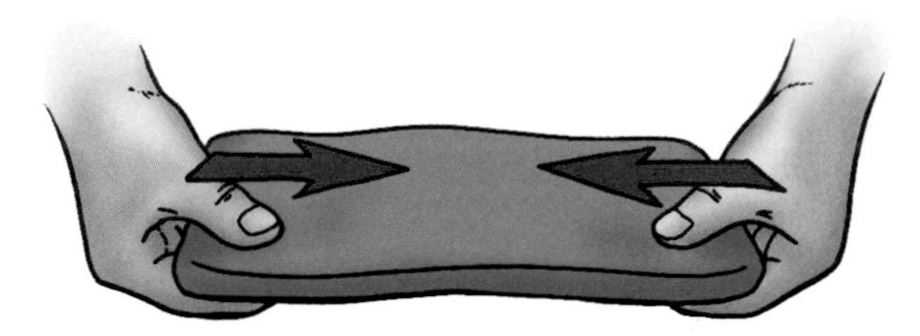

COMPRESS THE PAD BY PUSHING YOUR HANDS TOWARD EACH OTHER.
FIGURE **6.1**

SLIDE THE PADS PAST ONE ANOTHER.
FIGURE **6.2**

D. Now try this with two pads of different thicknesses. Did your results change?

E. Now work over a 10-cm opening between two desks or thick books. Place two thick pads side by side. Make the pads collide (push them together), as shown in Figure 6.3(A). What happens?

F. Place a thick and a thin pad side by side and make them collide. Again, work over a 10-cm opening between two desks or books if possible, as shown in Figure 6.3(B). With your group, discuss what happens. How does the behavior of the thick and thin pads differ from the behavior of the two thick pads? Why do you think this happens? Record your ideas and drawings.

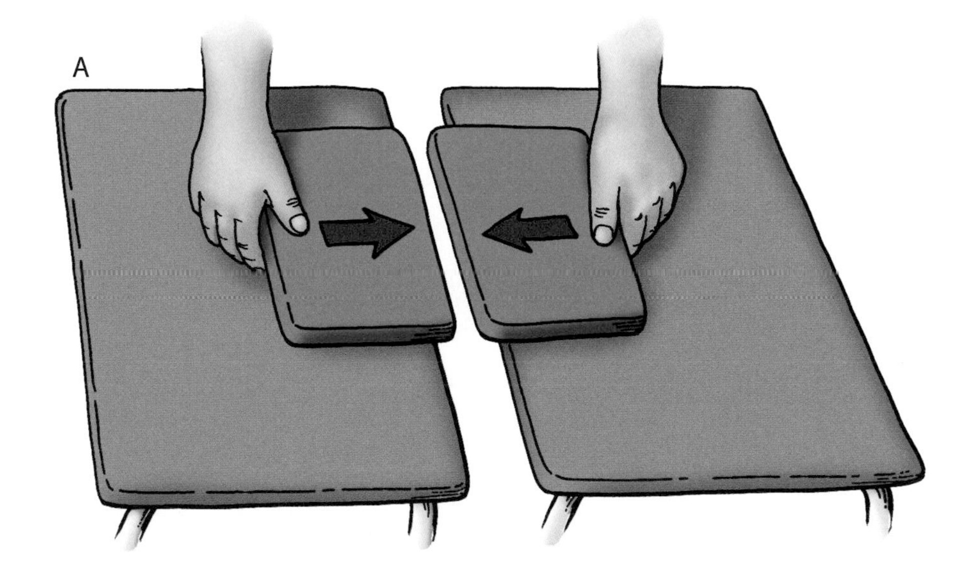

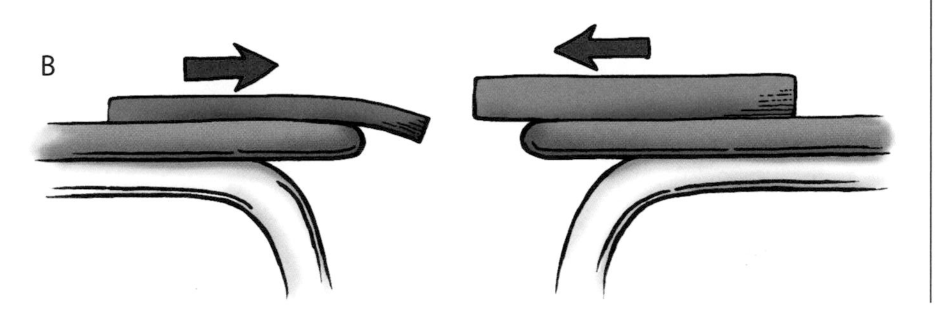

WORK OVER AN OPENING BETWEEN TWO DESKS OR THICK BOOKS. (A) MAKE TWO PADS OF EQUAL THICKNESS COLLIDE. (B) MAKE TWO PADS OF DIFFERENT THICKNESSES COLLIDE. WHAT HAPPENS TO THE PADS IN EACH SETUP?
FIGURE **6.3**

Inquiry 6.1 continued

4. If possible, examine a relief globe. Look for evidence of trenches and mountains. Then discuss with your group how you think they formed. (Think about the behavior of your pads and what you learned about plate movement from the CD-ROM.)

5. Observe as your teacher demonstrates the use of the Moving Plates Model™ as shown in Figure 6.4. Notice, the belts on top of the lid are black, with white just about to emerge from the slit in the center. We cannot see into the earth. This model will help you visualize the earth in three dimensions. Observe what happens when your teacher turns the knobs in opposite directions. Discuss the answers to the questions in each procedure that follows.

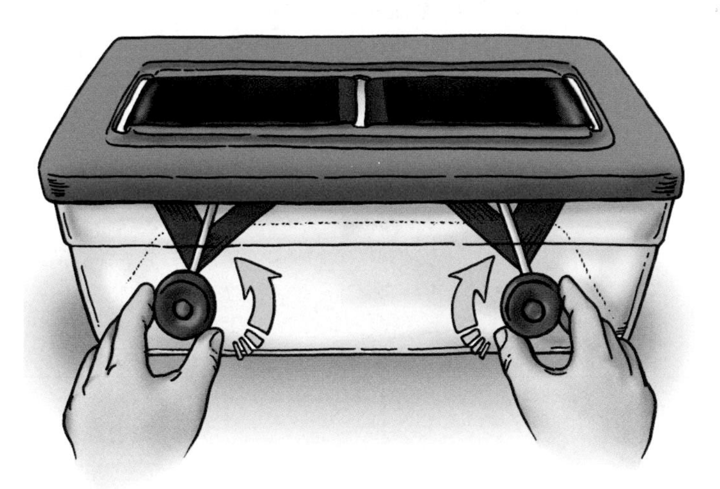

DEMONSTRATING THE MOVING PLATES MODEL™
FIGURE **6.4**

6 Your teacher has reset the Moving Plates Model by turning the knobs so that only black belts show on top and white belts are directly below the slit. Watch as your teacher very slowly turns the knobs so that both belts move away from the slit in the center.

A. How are the belts changing on either side of the center slit?

B. What happens to the black belts as they get near the edges of the lid?

7 Look at the dough continents—North America, Africa, and South America.

A. What observations can you make about their shapes?

B. Look on a map and discuss with your class what landforms exist on these continents and why you think they are there.

8 With the belts reset again, watch as your teacher tries different combinations of continents on the Moving Plates Model.

A. Which continents seem to fit well together?

B. Why do you think this is so?

9 South America and Africa fit together like puzzle pieces on the black belts and over the center slit, as shown in Figure 6.5. Watch as your teacher slowly turns the knobs so that the continents move away from one another. The current position of the continents is represented when the dough pieces are about 8 cm apart. This represents movement of the continents over millions of years.

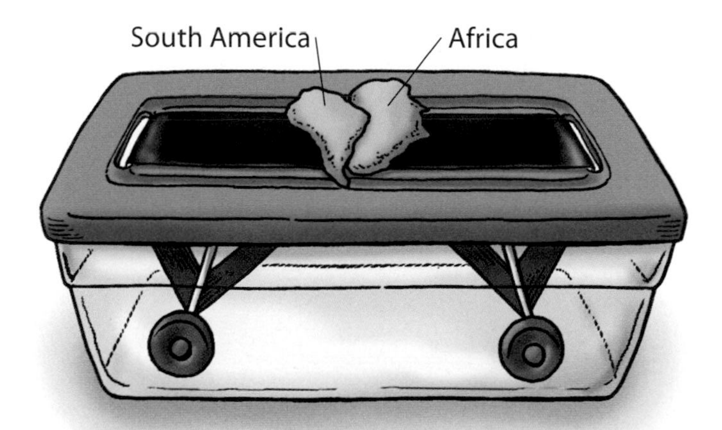

THE DOUGH MODELS FIT TOGETHER LIKE TWO PUZZLE PIECES. WHY DO THE CONTINENTS SEEM TO FIT TOGETHER?
FIGURE **6.5**

Inquiry 6.1 continued

10 Look at your Plate Tectonics World Map. What patterns do you observe in the shapes of the continents and the shape of the ridge in the center of the Atlantic? What do you think is responsible for these patterns? How do you think the Atlantic Ocean formed? Discuss your observations and ideas with your class.

11 Watch as your teacher reverses the knobs. North America and Africa start on the outer ends of the belts, as shown in Figure 6.6. Now you will go back in geological history. What happens as the continents collide?

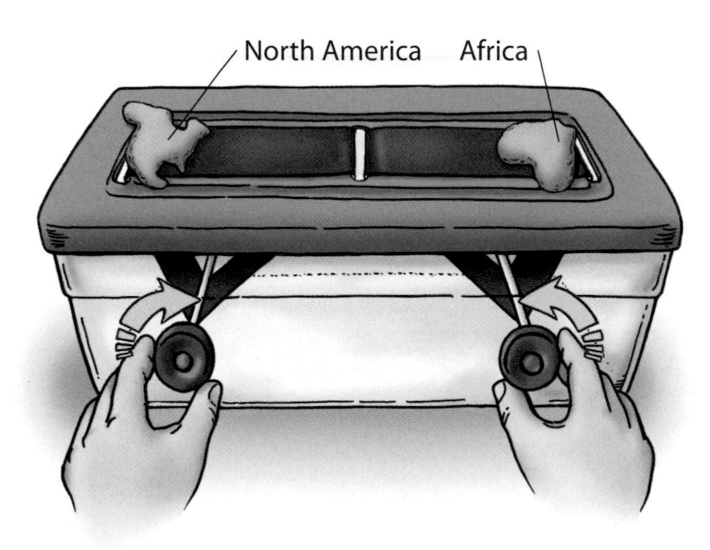

WATCH AS THE MODEL CONTINENTS OF NORTH AMERICA AND AFRICA MOVE TOWARD ONE ANOTHER.
FIGURE **6.6**

12 On your Plate Tectonics World Map, look at North America's eastern side. What landforms do you notice there? What process do you think might have been responsible for forming the mountain range there (called the Appalachian Mountains)?

13 Look at the relief globe. Do the following with your group, and discuss the answers to the questions as you work:

A. Feel the globe. What do you observe? How is it different from other globes you have used?

B. Where on the globe do you see evidence that plates collided in the past? What evidence do you have that plates may have collided there? Do you think those plates were continental, oceanic, or both? Why?

C. Where on the globe do you think plates are separating? Find evidence of this both within continents and under the ocean.

D. Find Japan and the Japan Trench. Feel these areas on the relief globe. What do you observe about these areas? What do you think is happening to the two plates that meet at this trench?

E. Find the middle of the Atlantic Ocean called the Mid-Atlantic Ridge. Feel this ridge on the relief globe. What do you observe about this landform? What do you think is happening to the two plates located along the Mid-Atlantic Ridge? Why do you think this ridge is higher than the rest of the ocean floor?

F. Are there other places on the globe similar to the Japan Trench and the Mid-Atlantic Ridge?

REFLECTING ON WHAT YOU'VE DONE

1. Answer these questions in your science notebook:

   A. How did the pads behave when pulled from opposite ends?

   B. How did the pads behave when compressed?

   C. If oceanic plates are colder, denser, and thinner than continental plates are, which pad do you think represented oceanic plates? Which pad represented continental plates?

   D. How did the density of the pads affect the way they behaved when you made them collide? Remember to state which type of plate, oceanic or continental, is involved in your answer.

   E. When do colliding plates on the earth form mountains?

   F. When do colliding plates form trenches?

   G. Why would a more dense oceanic plate slide under a less dense continental plate?

   H. Can plates ever move without forming new land? If so, when?

   I. How do you think colliding plates on the earth cause earthquakes?

2. Read "Colliding, Sliding, and Separating Plates" on pages 84–85.

3. By observing the Moving Plates Model™, what did you learn about the earth's plate boundaries? Answer the following questions in your science notebook, then share your responses with the class.

   A. How do you think the Moving Plates Model shows what happens on the earth when two plates separate?

   B. What causes the ocean floor to separate and "grow"?

   C. Think about what happened to the black belt as it reached the edges of the model's lid. What landform is created when the ocean floor sinks back into the earth?

   D. What patterns did you observe in the shapes of Africa and South America? How did the shapes of these continents compare with the shape of the Mid-Atlantic Ridge?

   E. What landform is created when plates separate? Give an example.

   F. What landform is created when two continental plates collide? Give an example.

4. With your class, look at the Earth's Fractured Surface wall map. Why do you think there are mountains north of India?

5. Read "Earth's Moving Plates: A Look Back" on pages 86–89.

READING SELECTION

BUILDING YOUR UNDERSTANDING

COLLIDING, SLIDING, AND SEPARATING PLATES

The crust of the earth, along with the rigid uppermost part of the mantle, is called the lithosphere. The lithosphere is 18 to 120 kilometers thick. It covers the earth's interior and is broken into pieces called plates. The rocks that make up these plates grind, collide, move past one another, and separate as they float on a flowing, taffy-like, solid upper mantle called the asthenosphere. The place where plates meet is called a plate boundary. At some plate boundaries, the plates collide, and mountains and trenches form, as shown in the photograph below. At other plate boundaries, the plates try to slide past one another. When this happens, energy builds up in the rock as it compresses or twists. When the force between the plates gets too great, the rock breaks, or ruptures, and an earthquake may occur.

CAN YOU SEE EVIDENCE IN THIS PHOTO OF MT. EVEREST AND NEIGHBORING MOUNTAINS THAT PLATES HAVE COLLIDED?

PHOTO: Astronaut photograph ISS008-E-13304 taken from the International Space Station on January 28, 2004. Image provided by the Earth Observations Laboratory, Johnson Space Center.

Lithospheric plates can be either continental or oceanic. Continental plates contain the earth's continents. They are thick but less dense than oceanic plates. Oceanic plates, which occur under the world's oceans, are thin and dense because of their composition.

More often, when two continental plates collide at what is called a convergent plate boundary, their edges crumple (imagine crumpling paper) and uplift to form mountains. This is what happened when the Indian-Australian Plate collided with the Eurasian Plate millions of years ago. Their collision formed (and is still forming) the Himalayas.

When a continental plate collides with an oceanic plate, it's a different story. The dense oceanic plate sinks and slides under the continental plate in a process called subduction, as shown in the top illustration at right. An older, colder oceanic plate can also slide under a younger, warmer oceanic plate, and the old plate moves deep into the earth. (This is because cold things are often denser than warm things.) When an oceanic plate moves under another plate, the bending of the sinking plate creates a trench, or deep valley, in the ocean floor. As the oceanic plate sinks deeper into the earth's hot interior at the trench, it is subjected to heat and pressure. As the temperature rises, part of the rock may melt. This molten, or melted, rock rises to the earth's surface and is the source of volcanoes.

Sometimes a new boundary can form under a continent. Hot rock flowing under a plate can cause the surface to thin and break. Then, the continent may split in two. The place where the plate splits is called a divergent plate boundary because the plates separate as melted rock flows to the surface. Like hot air that rises above cold air, melted rock inside the earth rises because it is less dense than the cooler, solid rock around it. The rising,

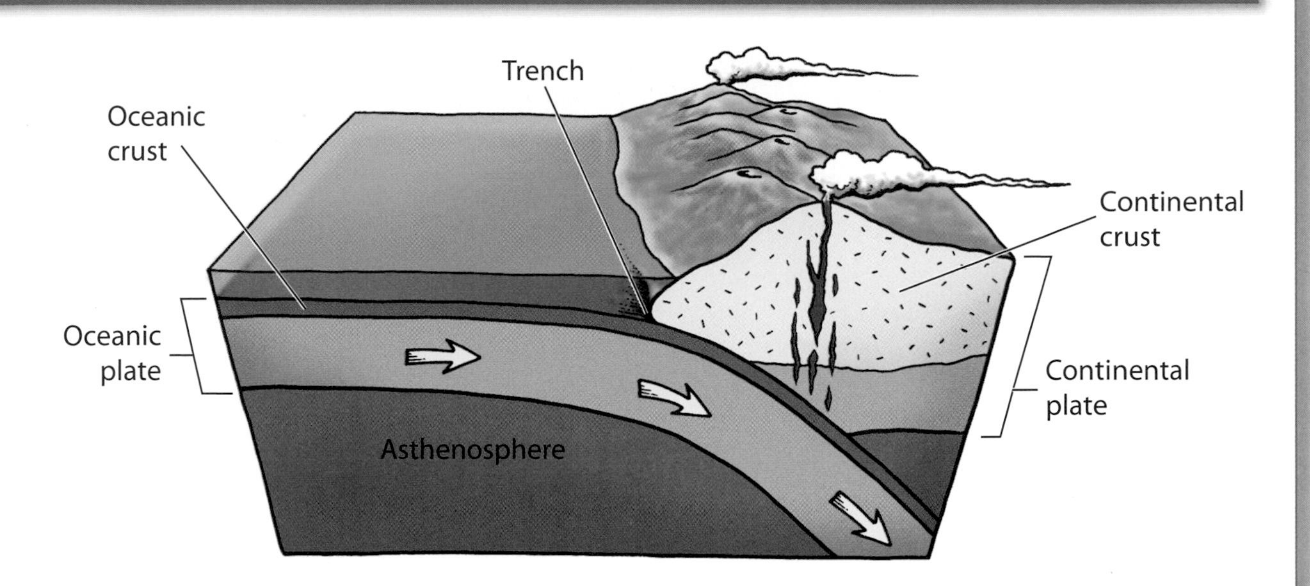

melted rock heats the crust and causes it to expand and bulge upward to form a ridge, or mountain-like landform. If water from an ocean enters the place where the plate split, a mid-ocean ridge, as shown in the illustration below, is formed.

Sometimes, at what is called a transform plate boundary, two plates slide past one another in such a way that it does not result in a new landform. You will learn more about these different types of plate boundaries in Lesson 7. ■

WHEN OCEANIC AND CONTINENTAL PLATES COLLIDE, THE OCEANIC PLATE SLIDES UNDER THE LESS DENSE CONTINENTAL PLATE. THE ANDES MOUNTAINS IN SOUTH AMERICA FORMED THIS WAY.

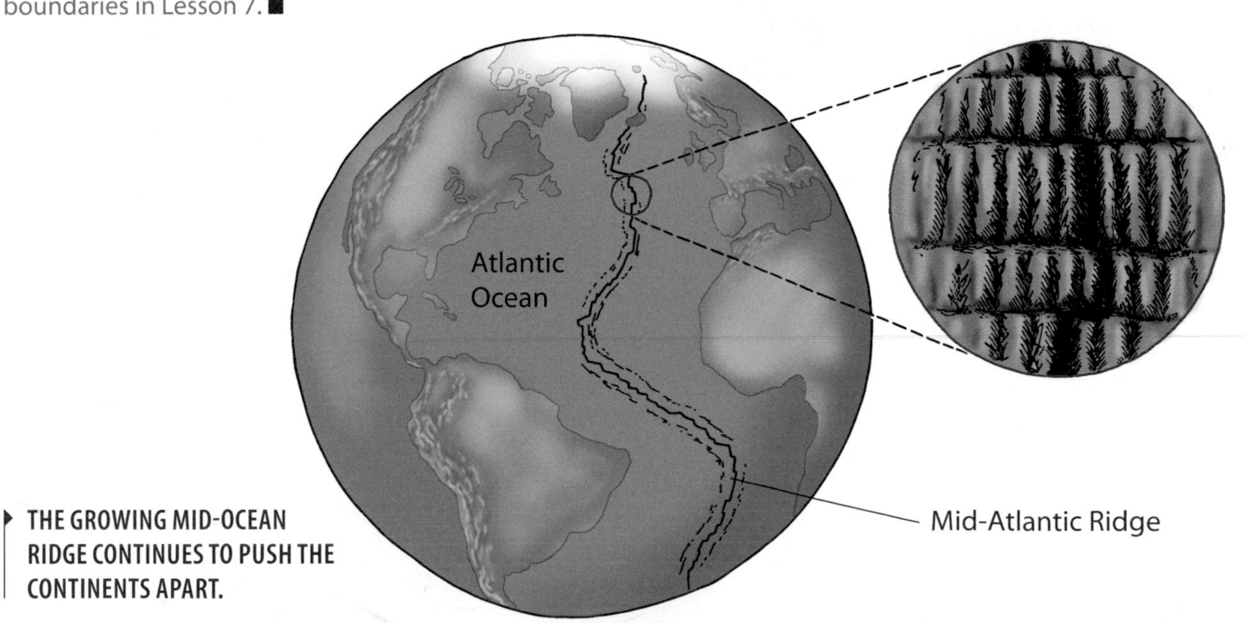

THE GROWING MID-OCEAN RIDGE CONTINUES TO PUSH THE CONTINENTS APART.

READING SELECTION

EXTENDING YOUR KNOWLEDGE

Earth's Moving Plates: A Look Back

You've observed that Africa and South America seem to fit together like pieces of a puzzle. Scientists now believe that all the continents were once a single landmass. Earth's hot mantle separated these continents over time, and oceans formed between them. How did scientists long ago explain these changes? Let's take a look.

LOOKING FOR EVIDENCE

For centuries, some scholars hypothesized that continents move. For example, in 1620, Sir Francis Bacon, an English philosopher, noticed that continental margins looked as if they would fit together.

In the mid-1800s, Antonio Pellegrini, a geologist, noticed that identical fossils were found on continents separated by wide oceans. He thought a great flood caused these oceans to form, separating the continents and their fossils.

Edward Suess, an Austrian geologist in the mid to late 1800s, claimed that the scratches and gouges from glaciers line up along the boundaries of separated continents. He also noted similarities among plant fossils on different continents. He hypothesized that these fossil similarities were evidence that long land bridges had once connected the landmasses. He believed that the bridges later sank beneath the ocean.

SIR FRANCIS BACON, ENGLISH STATESMAN AND PHILOSOPHER

PHOTO: Courtesy of Smithsonian Institution Libraries, Dibner Library of the History of Science and Technology, Washington, DC

In 1910, American geologist Frank B. Taylor explained that mountain ranges on distant continents line up. He theorized that large polar continents had broken apart, drifted toward the equator, and stayed there as a result of gigantic tidal forces. These forces, according to Taylor, were generated by the pull of gravity when the earth "captured" the moon.

THE BREAKUP OF A SUPERCONTINENT

After many scientists had gathered evidence, Alfred Wegener, a German meteorologist, proposed in 1912 the theory of continental drift. According to this theory, the continents were once united in one "supercontinent." Wegener named this continent Pangaea. He claimed that, over time, Pangaea had broken into pieces that drifted apart. South America and Africa had moved away from each other. North America and Europe had separated. His theory was supported by evidence from many different fields of science.

Wegener explained why the shorelines of different continents seem to match. He noted that mountain ranges of similar age and structure were now located on separated continents. Fossil animals, such as Mesosaurus (a freshwater reptile about 1 meter long), were found in countries on two different continents, Africa and South America. Because Mesosaurus lived only in fresh water, Wegener surmised that the continents at one time were connected. Finally, Wegener found evidence that continents currently in the Tropics were once covered with glaciers. This means the continents must have "drifted" or moved somehow.

Wegener proved that continents appeared to move over time, but he could not prove why these

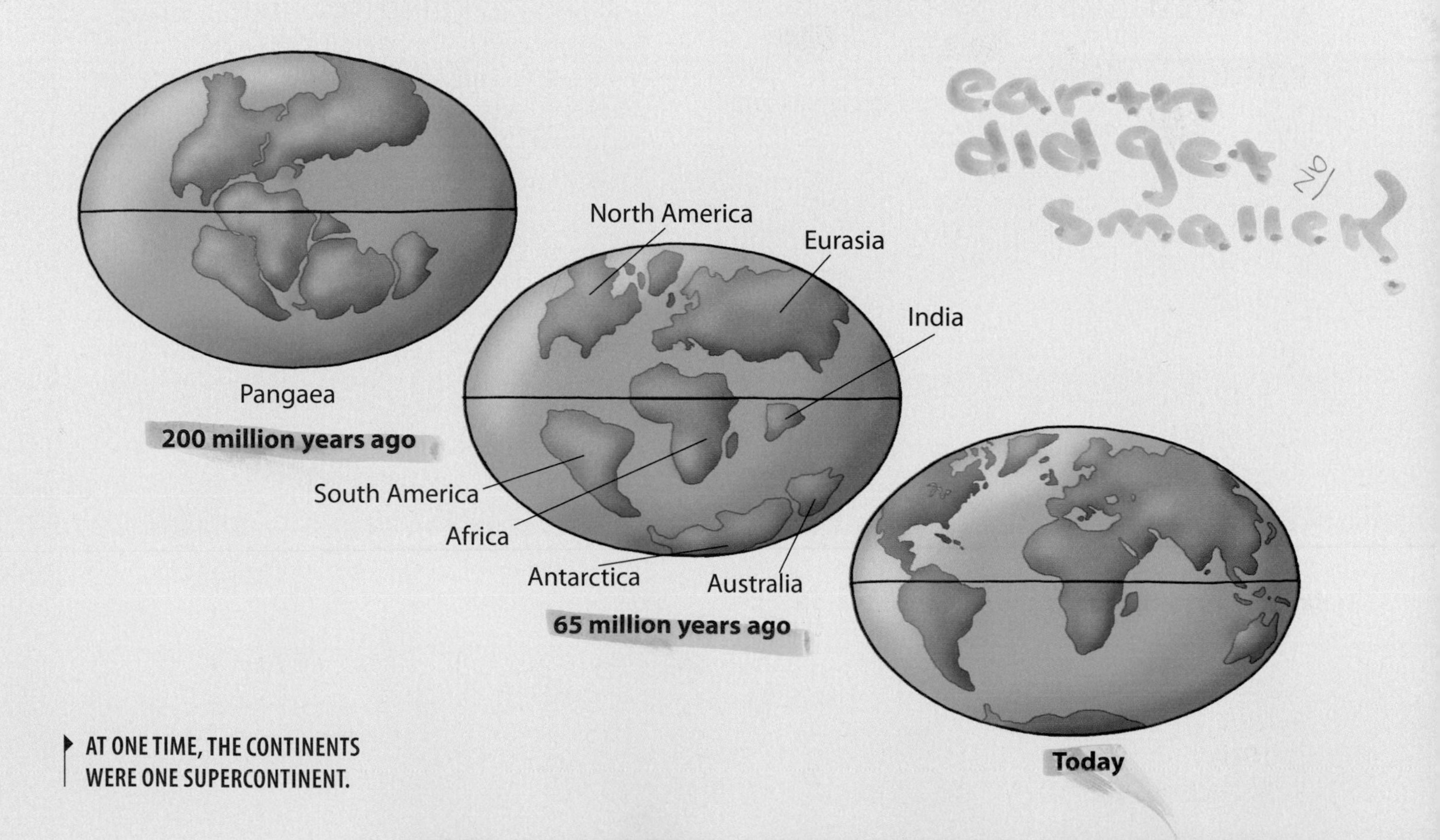

AT ONE TIME, THE CONTINENTS WERE ONE SUPERCONTINENT.

READING SELECTION

EXTENDING YOUR KNOWLEDGE

events took place. What caused the continents to drift? Wegener thought he had answers. He argued that both gravitational tidal forces and the earth's rotation were responsible for continental drift. However, it was fairly easy for scientists to prove that these forces were much too weak. His explanations were not accepted.

SEAFLOOR SPREADING AND THE DRIFTING CONTINENTS

In the late 1950s and early 1960s, years after Wegener's death, new data about the seafloor emerged that enabled geologists to suggest why continents appear to drift. Evidence from seafloor fossils and magnetic data suggested that younger parts of the floor were located closer to the mid-ocean ridge, while older parts of the seafloor were farther from the mid-ocean ridge, near the trenches. These data prompted the theory of seafloor spreading, which states that a force within the earth drives the ocean floor apart and allows new oceanic crust to form.

PLATE TECTONICS

In the late 1960s, scientists combined information on seafloor spreading and continental drift to propose the plate tectonics theory. This theory states that rigid plates move away from mid-ocean ridges, where new lithosphere is constantly being formed. It also proposed that old lithosphere moves away from these ridges and toward ocean trenches. At the trenches, old ocean lithosphere sinks into the earth. The plate tectonics theory also states that mountain chains of volcanic islands, such as Japan, form along trenches, where events such as earthquakes and volcanoes occur.

This theory was a major breakthrough for scientists. For many years, they had known that the earth's surface was slowly drifting, but they couldn't explain why. Today, they think they have found the answer. But there is still much to be learned about the earth's hidden interior and its effects on the planet's ever-changing surface. ■

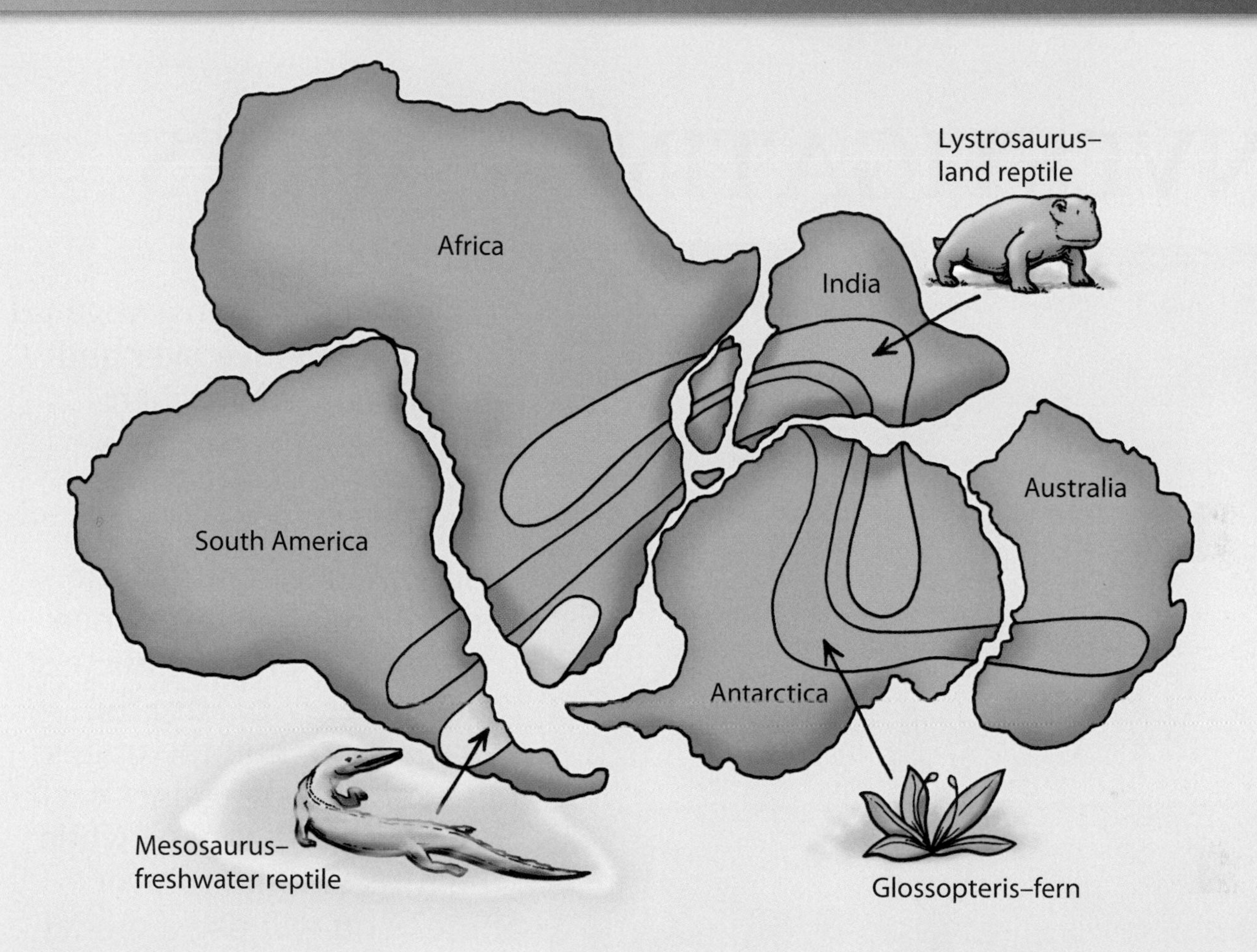

WEGENER FOUND THAT EVIDENCE FROM FOSSILS, FOR EXAMPLE, SUPPORTED HIS THEORY THAT PLATES MOVE.

DISCUSSION QUESTIONS

1. Why did it take so long for scientists to agree that continents moved on the earth?
2. What evidence supports the theory of continental drift?

LESSON 7

INVESTIGATING FAULTS

EVIDENCE OF THE HAYWARD FAULT IN CALIFORNIA

PHOTO: Naotake Murayama/creativecommons.org

INTRODUCTION

In Lesson 6, you investigated how plates move over and under each other where they meet and how they can spread apart on the ocean floor. In this lesson you will investigate in more depth what happens when continental plates slide past each other. What are the forces like when the plates slide and push against each other? What happens when the forces build up over time? You will use a model of a fault to investigate these questions and see how the forces created by plate movement cause rock to rupture and release vast amounts of energy in the form of an earthquake.

OBJECTIVES FOR THIS LESSON

- Classify materials as brittle or ductile.
- Design an investigation.
- Investigate the effects of applying a force to a model of a fault.
- Relate the interaction of forces at boundaries to the occurrence of earthquakes.

MATERIALS FOR LESSON 7

For you

1	copy of Student Sheet 7.1: Investigating Faults: Recording and Analyzing Data
1	pair of safety goggles

For your group

1	set of miscellaneous brittle materials
1	set of miscellaneous ductile materials
1	Fault Laboratory™
1	spring scale
4	plastic centimeter cubes
1–2	wooden tongue depressors
1	strip of masking tape
1	ruler

GETTING STARTED

1. With your class, discuss what happens to rock when plates move.

2. Predict what would happen if you applied a force to a stick at both ends. Watch as your teacher tests your predictions.

3. Obtain a set of brittle and ductile materials for your group. Observe their properties. Then classify each item as brittle (breaks when a force is applied to it), ductile (bends, stretches, or flows when a force is applied), or both. Record your group's classifications in your science notebook using either a list or Venn diagram. What other items could you put in these categories? List them.

FOLDED GNEISS
FIGURE **7.1**
PHOTO: U.S. Geological Survey

4. Discuss your group's classifications and reasons for choosing each with the class. Answer the following:

   A. Which materials were brittle? Which were ductile? Why did you classify them this way?

   B. How did you apply force to each object? How did each object respond to that force?

   C. Did temperature and pressure affect the behavior of your objects? If so, how?

   D. Based on these observations, what do you think are the conditions that affect how an object responds to a force?

5. The rock that the earth's plates are made of responds to the forces caused by plate movement in ways that are similar to the ways the ductile and brittle objects responded to force. Examine the rock shown in Figure 7.1. How do you think this rock became folded?

6. Think about what you have learned about brittle and ductile objects. How do you think this relates to rocks in the earth? Discuss the following questions with your partner and be prepared to share your answers with the class:

   A. In what part of the earth might rocks be more brittle and fracture more easily?

   B. How do you think the tendency of rocks to fracture relates to earthquakes?

READING SELECTION

BUILDING YOUR UNDERSTANDING

EARTHQUAKES AND FAULTS

Push your hands together hard. If your palms are flat against each other, not much will happen. But if you push your hands together at an angle, they will slide past each other. The earth's plates work something like this. Most earthquakes are the result of huge pieces of rock in the earth that rub or press against each other as a result of changes inside the earth. As forces are applied to the rock, the rock slightly deforms. Energy builds up within and between pieces of rock. Suddenly, they slip past one another. Energy is released, and the ground ruptures and shakes. The longer the force is applied to the rock, the greater the amount of energy that will be stored in the rock, and the more severe the earthquake.

Plate boundaries occur along fractures or breaks in the earth's outer layer. A fault is a fracture along which blocks of rock on opposite sides of the fracture move. One type of fault is shown in the photo at right.

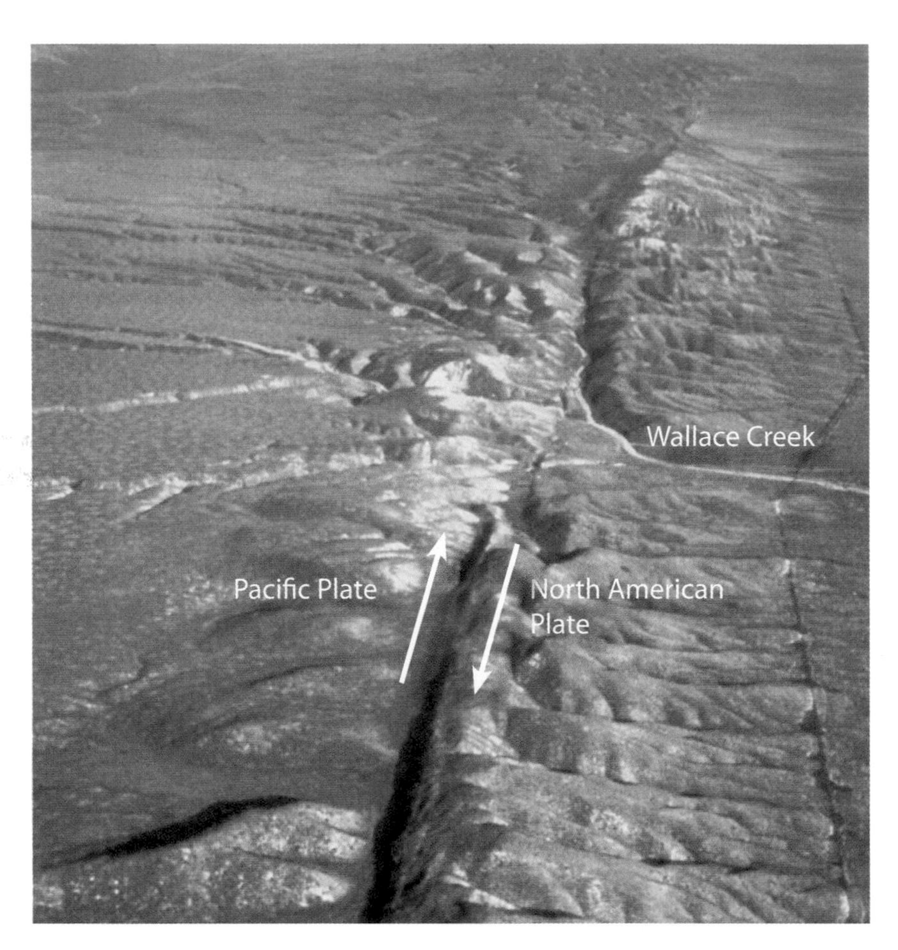

ALONG THE SAN ANDREAS FAULT, PLATES SLIDE PAST ONE ANOTHER. THE PATH OF WALLACE CREEK, WHICH FLOWS ACROSS THE FAULT, HAS CHANGED BECAUSE OF MOVEMENT ALONG THE FAULT. NOTICE THE RIDGES, WHICH HAVE BEEN FORMED BY HUNDREDS OF MOVEMENTS ALONG THE FAULT.

PHOTO: R.E. Wallace/USGS/NGDC/NOAA, Boulder, CO

TRANSFORM FAULTS

The San Andreas Fault in California, shown in the photo, is an example of a transform fault. The fault marks the line where the North American Plate and the Pacific Plate move horizontally past each other. No new landform is created along this boundary, but the rocks along the fault have jagged surfaces that hook and catch on each other, creating friction. The force behind each moving plate drags these hooked surfaces powerfully. If they are dragged strongly enough, the fault breaks, bending and fracturing the rock, and allowing the plates to slip past each other in opposite directions. On the surface, we experience this as a shallow earthquake. Fences, rivers, and other structures on the land along the fault can change shape as a result of the forces created by the moving plates.

One variable that affects the crust's behavior along a fault is the amount of friction between fault surfaces. Young faults have rough surfaces, but when a fault ruptures repeatedly, its rough surfaces, or protrusions, wear down and become smooth.

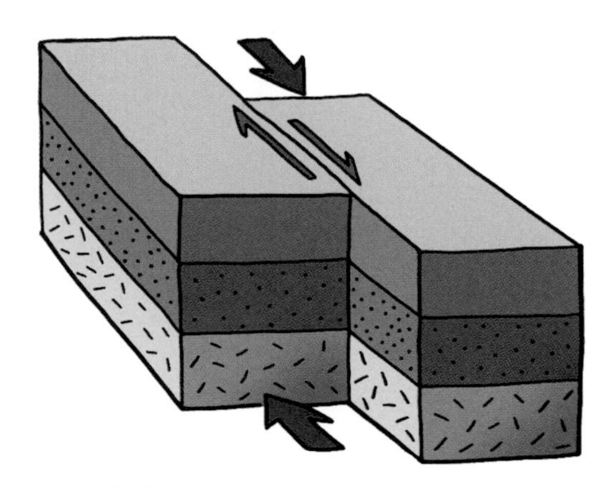

PLATES SLIDE PAST ONE ANOTHER ALONG THIS TRANSFORM FAULT.

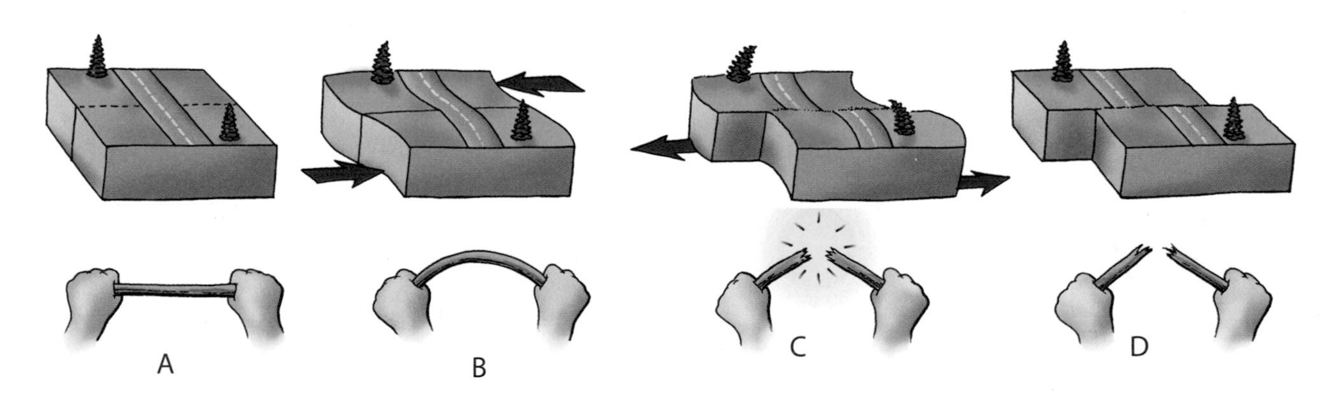

IF YOU APPLY FORCE TO A STICK, IT BENDS FIRST, THEN BREAKS. APPLY FORCE TO A ROCK, AND THE ROCK WILL STORE ENERGY. WHEN THE STRENGTH OF THE FORCE BECOMES GREATER THAN THE STRENGTH OF THE ROCK, THE ROCK WILL BREAK, AND AN EARTHQUAKE WILL OCCUR.

READING SELECTION

BUILDING YOUR UNDERSTANDING

EVIDENCE OF THE SUBDUCTING DENALI FAULT IN ALASKA CAN BE FOUND IN THIS TRACE ALONG A GLACIER (NOTE HOW THE GROUND TO THE RIGHT OF THE TRACE IS HIGHER THAN THE GROUND TO THE LEFT).

PHOTO: U.S. Geological Survey

SUBDUCTION FAULTS

Another type of fault forms along a subduction zone. The earthquakes on these faults can be extremely large and destructive. The Alaska quake that you studied in Lesson 3 happened on a subduction fault. In this type of fault, seafloor slides beneath the continental plate and is melted in the intense heat below. Again, there is friction between the two plates, but the friction is generated between the surface of the lower plate and the edge and bottom of the top plate. The west coast of Washington and Oregon also experience subduction plate quakes. There, the Juan de Fuca Plate slides under the North American Plate. The beautiful, volcanic Cascade Mountain range developed as a result of the forces from this collision.

AN EARTHQUAKE IN SAUDI ARABIA IN 2009 CAUSED THIS RUPTURE ON THE GROUND SURFACE FROM THE SPREADING OF THE PLATES BENEATH. NOTE THAT THE RUPTURE IS LARGER IN THE SOFT SEDIMENT (FOREGROUND) THAN IN THE ROCKY AREA (BACKGROUND).

PHOTO: U.S. Geological Survey/photo by John Pallister

SPREADING FAULTS

A third major type of fault happens along ocean bottoms where two plates pull apart and magma wells up into the chasm, causing seafloor spreading and the formation of new ocean floor. ■

INQUIRY 7.1

INVESTIGATING FAULTS WITH MODELS

PROCEDURE

1. Discuss with your group what you already know about faults. What questions do you have about faults? Record your group's ideas in two lists in your science notebook. Title the lists. Share your ideas with the class.

2. Look at the Fault Laboratory™. You will use this model to investigate movement along a fault where blocks of rock slide past one another. What do you think will happen when a force is applied to the blocks? Discuss your hypothesis with the class.

3. If you have not read "Earthquakes and Faults" on pages 94-97, read it now.

4. Preview Steps 6 through 12 of the Procedure with your teacher.

5. Discuss with the class how you will record your data. If you graph your data, how will you graph it? Use Student Sheet 7.1: Investigating Faults: Recording and Analyzing Data to record and analyze the data you will collect as you experiment with the blocks. Note that you will conduct trials with no Velcro® strips, and with one, two, and three strips.

SAFETY TIP

Wear safety goggles at all times when working with the Fault Laboratory. The cord might break or slip off the hook unexpectedly.

6 Collect your materials. Use Figure 7.2 to set up your Fault Laboratory. Check that each of the following steps is completed:

A. Carefully place the block with the hole so that the long strip of soft, looped Velcro is facing the "fault." Use the bolt, washers, and wing nut to secure the block with the hole to the box.

B. Notice how each side of the block has a different number of strips of "hooked" Velcro. For each test, you will rotate the block so there is more hooked Velcro (or frictional resistance) between the two blocks.

C. Slide the solid block (with no hole) in place next to the fixed block. Make certain the soft loop Velcro on both blocks is touching and secure. Use one to two tongue depressors as spacers between the plastic box and the solid, unfixed block only if the blocks are loose. This will push the blocks together.

D. Make a loop or tightly knot the cord to the hook on the solid, unfixed block.

E. Thread the cord through the hole opposite the hook.

F. Use a loop or tightly knot the cord to attach the spring scale.

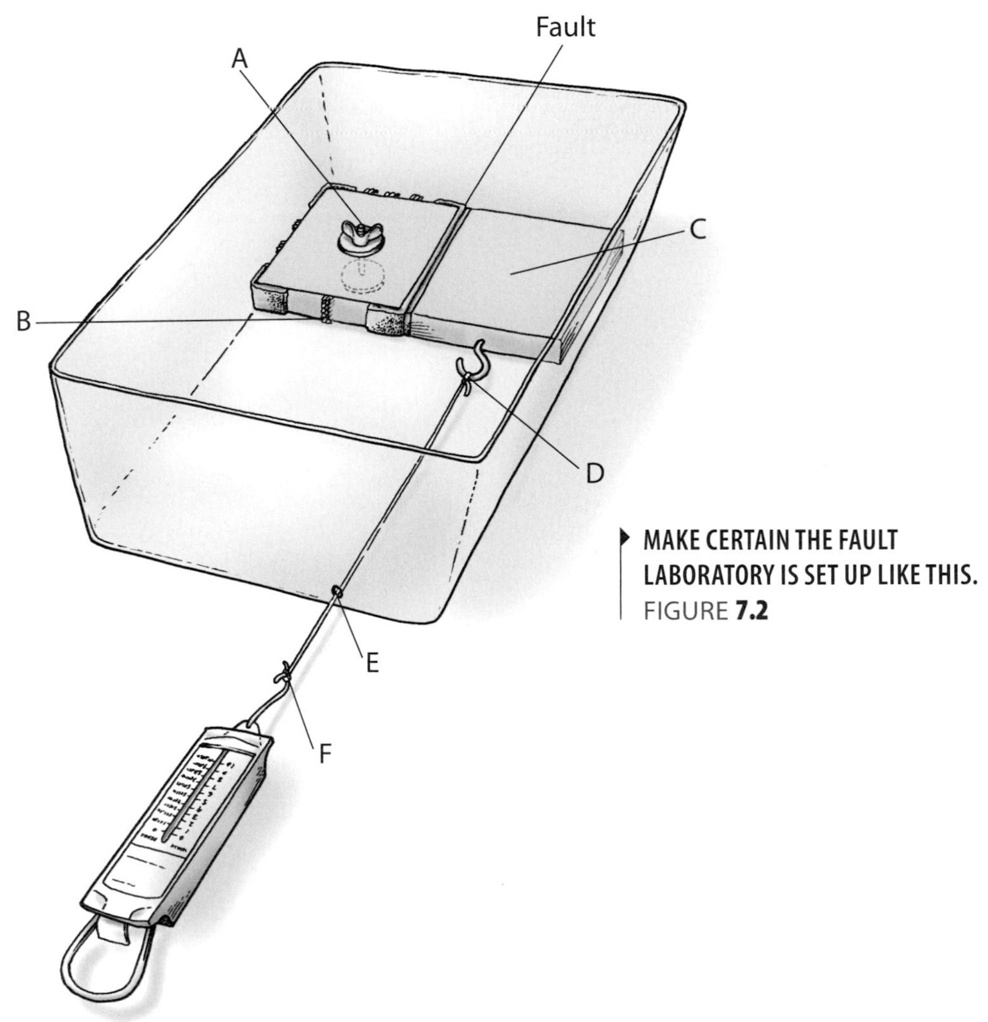

MAKE CERTAIN THE FAULT LABORATORY IS SET UP LIKE THIS.
FIGURE **7.2**

Inquiry 7.1 continued

7. Experiment freely with the unfixed block. How does the block move when you pull it? Try putting masking tape across the boundary, or "fault," where the two blocks meet. Pull on the block again. What happens to the tape that indicates stress is building along the fault? Reset the blocks. Place houses (plastic centimeter cubes) on the blocks along the fault. Pull on the block again.

8. Remove the tape and houses. Put them aside for now. Collect and analyze data to find out how the amount of frictional resistance along a fault affects the way it ruptures. Follow Procedure Steps 9 through 12 as you work.

9. Pull on the spring scale, which will apply a sliding force to the unfixed block. How much force do you have to apply to the block before the fault ruptures (that is, before the block moves abruptly)? Record the maximum force in your data table for "0," since you are not using any hooked Velcro® on the blocks, and there is very little frictional resistance along your fault. Conduct three trials.

10. Now rotate the unfixed block so there is one strip of hooked Velcro along the fault. Repeat Procedure Step 11. Conduct three trials. Record your force data each time next to "1" on Table 1 on Student Sheet 7.1.

11. Now repeat these procedures with two strips of hooked Velcro, recording data next to "2" in Table 1 as you do so.

12. Finally, rotate the block again and test the force using three strips of hooked Velcro. Record your data next to "3" in Table 1. Place houses (plastic centimeter cubes) on the blocks. What happens to the houses during an "earthquake"?

13. Complete Student Sheet 7.1 by calculating the averages and plotting the data.

14. Clean up.

THE WRINKLES IN THIS ROAD ARE A SIGN THAT PRESSURE IS BUILDING ALONG A FAULT. GEOLOGISTS CALL THIS GRADUAL SLIPPING "CREEP."

PHOTO: Sue Hirschfeld/National Geophysical Data Center/NOAA, Boulder, CO

FIGURE **7.3**

REFLECTING ON WHAT YOU'VE DONE

1. Draw conclusions about faults and earthquakes on the basis of your results. Answer these questions:

   A. How did the amount of friction along the fault affect the amount of force needed to rupture the fault? Use data to support your answer.

   B. Under what conditions did the blocks rupture more abruptly?

   C. Under what conditions did the block slip (move slowly) but not rupture?

   D. Under what conditions were the biggest earthquakes produced? Use data to support your answer.

   E. Think about what happened with the masking tape. Is there any sign on the earth's surface that the earth is moving slowly beneath the crust? (Look at Figure 7.3 and use the caption to answer this question.)

2. How would you define the term "fault"? Write your definition in your science notebook.

3. With your class, use the Earth's Fractured Surface wall map and find one place on the earth where a fault is located.

4. Watch the video *Earthquakes*.

LESSON 8

CONVECTION IN THE MANTLE

ARISTOTLE, A GREEK PHILOSOPHER WHO LIVED FROM 384 TO 322 BC

PHOTO: Courtesy of Smithsonian Institution Libraries, Dibner Library of the History of Science and Technology, Washington, DC

INTRODUCTION

Like many people in his day, Aristotle, an ancient Greek philosopher, tried to explain why earthquakes and volcanoes occurred. He believed fire burned deep within the earth. He thought that when winds from the atmosphere were drawn underground, they mixed with the flames inside the earth and then exploded upward toward the surface. The results, according to Aristotle, were earthquakes and volcanic blasts. More than 1000 years later, Benjamin Franklin thought earthquakes came from a spark in the ground.

But why *does* the ground rattle and shake? Both Aristotle's and Franklin's theories had one thing in common—heat. And although theories since their times have changed, the earth's internal heat remains the explanation for why the earth's plates move.

In this lesson, you will investigate convection in the earth's mantle. You have seen convection many times in your life. Think of smoke rising from a campfire, or the air above the hot blacktop. Maybe you have felt cold air as you descended into a cave, or the draft under your front door in the winter. These convection currents happen when air at different temperatures mixes.

You will apply what you know about convection to better understand the earth's mantle. How do convection currents in the mantle cause the earth's plates to separate and sink back into the earth? What causes the continents to move over time? Using a special fluid that is very sensitive to heat, you will model convection currents in the mantle. By viewing computer images, you will be able to see what happens inside the earth. You will then relate convection cells to the movement of the earth's plates.

OBJECTIVES FOR THIS LESSON

- Use a flow indicator, Carolina™ Convection Fluid, to model convection currents in the earth's mantle.
- View computer images of the earth's interior to observe convection in the mantle.
- Use appropriate vocabulary when communicating ideas about the earth's interior layers.
- Identify movement in the earth's mantle as one cause of plate movement, earthquakes, and volcanoes.

MATERIALS FOR LESSON 8

For you

1 copy of Student Sheet 8.1a: Convection in the Mantle
1 copy of Student Sheet 8.1b: Earthquakes Review
1 pair of safety goggles

For your group

1 jar of Carolina™ Convection Fluid, capped
1 tea candle
2 wooden blocks
1 flashlight
2 paper towels

GETTING STARTED

1. With your class, brainstorm what you know about why the earth's plates move.

2. Think about what you know about convection. Discuss these questions with your class:

   A. What do you know about convection in the air?

   B. What do you know about convection in the ocean?

   C. How do you think convection in the mantle might be related to plate movement?

3. Your teacher will demonstrate the Moving Plates Model™ used in Lesson 6. Watch as the belts on the model move. Describe to your class what is happening to the belts at the top of the model. Based on the Moving Plates Model, what do you think is causing the plates on the earth to rise and separate at the ridges and sink at the trenches? Discuss this question with the class.

WHAT DO YOU KNOW ABOUT CONVECTION IN THE OCEAN?

PHOTO: Randolph Femmer/life.nbii.gov

INQUIRY 8.1

MODELING CONVECTION IN THE MANTLE

PROCEDURE

1. Look at the materials your teacher has set out. Pick up one jar for your group. It contains a special fluid that is a flow indicator. Observe the fluid in the jar.

2. Share your observations with the class. Discuss how you might use it to observe convection cells.

3. Obtain one copy of Student Sheet 8.1a: Convection in the Mantle. Discuss it with your teacher. Review the Safety Tips with your class.

SAFETY TIPS

The fluid in the jar is nontoxic, but do not loosen or remove the cap on the jar.

Wear your safety goggles.

Be careful when working with an open flame.

If you have long sleeves, push them up. Never reach across the flame. If your hair is long, tie it back.

Be very careful when working with the heated jar.

Inquiry 8.1 continued

4 Collect and set up the rest of the materials for your inquiry.

A. Shake the Carolina™ Convection Fluid so that you can better observe its flow.

B. Set up your equipment as shown in Figure 8.1.

C. Call your teacher over to light your candle and turn out the lights.

D. Place the lit candle under the jar.

5 Shine the flashlight on the front of the jar to observe the movement of the fluid, as shown in Figure 8.2. Reposition the candle to observe different patterns of movement in the fluid.

6 Shine the flashlight down on the jar (on the glass surface parallel to your table or desk).

7 Add arrows to the diagram on Student Sheet 8.1a to depict the direction in which the fluid moves. Label the diagram to show where the warmer fluid is and where the cooler fluid is.

8 Discuss your observations with your group, and then with your class. Complete Part A of Student Sheet 8.1a by summarizing your observations of the fluid.

9 Clean up.

A. Blow out the candle.

B. Let the jar cool, then use a dry paper towel to wipe off any black carbon marks from the candle that might be on the jar.

C. Return the jar of Carolina™ Convection Fluid to the materials center. It will be used by other classes.

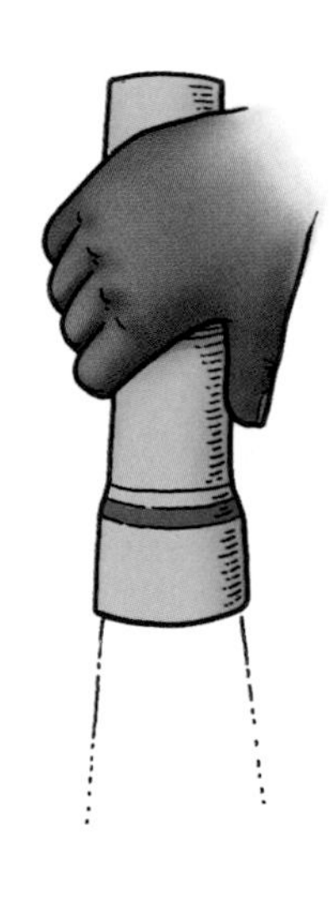

THE SETUP FOR THE CAROLINA™ CONVECTION FLUID
FIGURE **8.1**

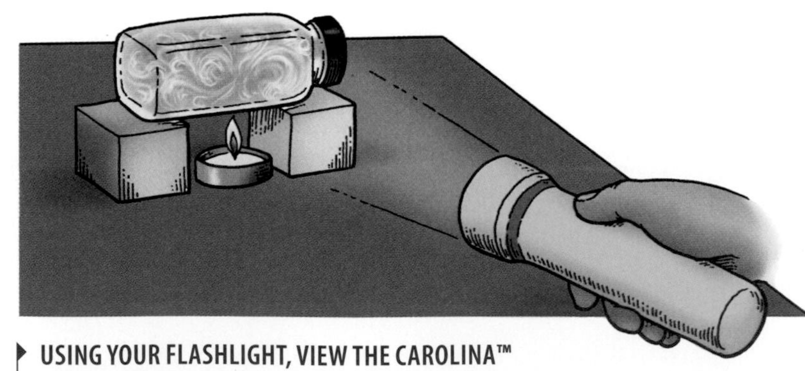

USING YOUR FLASHLIGHT, VIEW THE CAROLINA™ CONVECTION FLUID FROM THE FRONT AND TOP OF THE JAR.
FIGURE **8.2**

REFLECTING ON WHAT YOU'VE DONE

1. Discuss the answers to these questions with your group. Then discuss them with the class.

   A. What observations did you make of the heated fluid?

   B. Under what conditions could you observe convection cells forming inside the jar? How did they move? Compare this motion with what you observed using the Moving Plates Model™.

   C. What happened to the fluid near the upper surface of the jar (parallel to the table)?

2. Share with the class your responses on Part A of Student Sheet 8.1a.

3. Relate your observations of the jar and candle to the earth. Answer these questions in your science notebook:

   A. What causes convection currents in a gas or liquid? (Think back to the Moving Plates Model discussed during "Getting Started" of this lesson.)

   B. Based what you have seen in the jar, what effect do you think convection in the hot mantle might have on the earth's plates? What observations of the fluid inside the jar support your explanations?

4. Watch as your teacher shows you Segments 10 and 16 on the CD-ROM *The Theory of Plate Tectonics*.

5. Complete Part B of Student Sheet 8.1a by comparing your diagram of the jar with the diagram of the earth shown on the sheet. Label your diagram of the jar to show how it is a model of the earth's interior.

6. Look ahead to the lessons on volcanoes, in which you will study how the earth's internal heat forms volcanoes.

7. Review what you have learned about earthquakes and plate movement by completing Student Sheet 8.1b: Earthquakes Review.

READING SELECTION

EXTENDING YOUR KNOWLEDGE

Convection Is the Answer

When Alfred Wegener introduced his theory of continental drift to the scientific world in 1912, even his most sympathetic listeners had a hard time taking him seriously. For one thing, he was telling them that the earth's land masses were rooted in the mantle but drifted around the globe, gliding through the ocean floor like icebergs through the sea. He couldn't explain what caused the continents to move or how they managed their trips across the ocean floor.

Geologists found his ideas completely implausible. Instead, they favored the idea that changes in the crust, like the eruption of mountain ranges, were caused by temperature fluctuations in the earth, which caused the earth to expand or shrink. Mountain ranges, under that hypothesis, were a sort of bunching up of the crust over shrinking earth. Adding to their skepticism towards Wegener and his theory was the fact that Wegener wasn't a geologist at all; he was a meteorologist. However, several scientists following Wegner provided theories and evidence that, after decades, brought the scientific community around to accepting continental drift.

First, a British geologist named Sir Arthur Holmes proposed that the shifting, semi-liquid mantle under the earth's crust flowed in a regular pattern. Heat, given off by decaying radioactive elements in the earth's core, drove the mantle's flow. And, he said, the slowly flowing mantle dragged continents with it. "Slowly" here meant quite slowly indeed; at most, a continent might move 15 cm (about 6 inches) over an entire year.

Holmes' theory relied on an understanding of thermal

THE SUITCASES ON THIS CONVEYOR BELT ARE CARRIED ALONG LIKE CONTINENTAL PLATES ON A CIRCULATING MANTLE.

PHOTO: Dave Glass/creativecommons.org

convection. As part of a fluid heats, it grows less dense and rises. Imagine a heated portion of the mantle moving upward, then cooling, becoming denser, and sinking downward again, creating a circular current. Like airport carousels that move luggage, said Holmes' theory, circulating parts of the mantle moved the continents.

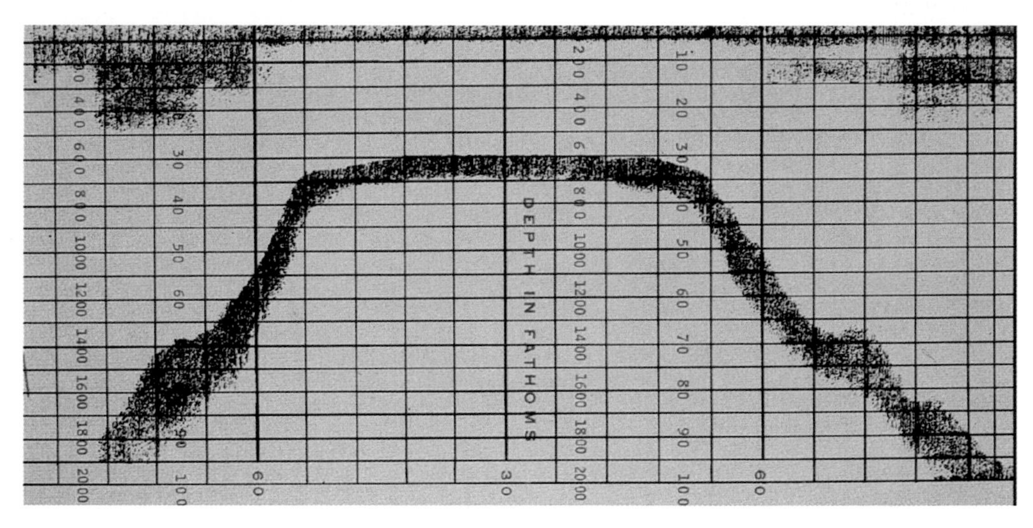

GRAPH OF A GUYOT (AN UNDERSEA MOUNTAIN FORMED FROM AN EXTINCT VOLCANO) DISCOVERED BY HARRY HESS IN THE 1940S

PHOTO: NOAA Central Library

Even though Holmes's theory provided a mechanism for continental drift, no one paid it much attention, and people were still not convinced. It took the work of five other scientists to generate a consensus in favor of the theory within the scientific community.

HARD DATA TO THE RESCUE

Working in the 1950s and 60s, Harry Hess and Robert Dietz studied the ocean floor and suggested that continental land masses weren't plowing through the seabed; instead, the seabeds were split at thin points along the crust, and magma came up through those splits, forcing parts of the ocean floor away from each other and forming new crust. In other words, Hess and Dietz figured that the continents weren't moving on their own; as chunks of seafloor slowly moved apart, they carried continents along with them. At the far edges of the crustal plates, the crust sank down into other rifts and melted back into the mantle. Hess and Dietz collected other data supporting the theory of mantle convection driving continental drift. For example, mid-oceanic ridges with volcanoes signaled that magma was welling up at plate boundaries.

DR. ROBERT DIETZ EXPLAINING HIS FINDINGS ON THE INDIAN OCEAN SEVERAL YEARS AFTER HE PROPOSED THAT SEAFLOOR SPREADING OCCURS AT ACTIVE OCEAN RIDGES.

PHOTO: Archival Photography by Steve Nicklas, NOS, NGS/NOAA

READING SELECTION

EXTENDING YOUR KNOWLEDGE

Soon after Hess and Dietz presented their ideas, three other geologists—Fred Vine, Drummond Matthews, and Lawrence Morley—suggested that new discoveries about the magnetism of freshly formed rocks would support their theory of seafloor spreading. Their theory turned out to be correct, and it was crucial to the scientific community's acceptance of continental drift as reality.

THE WORK'S NOT DONE

Today, we continue to seek information about the details of mantle convection and how exactly it contributes to continental drift. Where does convection occur, and what are the forces that generate and direct its flow? Keep in mind that the mantle's flow can be influenced by heating from the bottom, cooling at the top, and movement of crustal materials at plate boundaries. Because the forces of mantle convection are operating deep within the earth's interior, they are difficult to study. It is tricky to determine exactly how they contribute to the mantle's convection.

The "plume model" of mantle convection assumes that magma wells up from deep in the mantle and generates the convective currents that move the continental plates. Imagine a simmering bowl of thick soup with crackers floating on the top. As the cooler soup sinks down and the hotter soup rises, currents are generated that move the crackers around. So, while the mantle is very active, the continental plates in this model are passive, reacting to the activity in the mantle. The plume model has frequently been used to diagram and explain mantle convection.

WHAT WOULD HAPPEN TO THE CRACKERS IN THE SOUP IF YOU HEATED IT TO BOILING?

PHOTO: LWY/creativecommons.org

In recent years, however, evidence has accumulated for the "plate model" of mantle convection. This model gives the continental

APOLLINARIS PATERA IS AN ANCIENT VOLCANO ON MARS WITH A CALDERA THAT IS 80 KILOMETERS (50 MILES) ACROSS!

PHOTO: NASA/JPL/Malin Space Science Systems

plates themselves an active role in generating the motion of the mantle below. According to this model, continental plates affect the mantle's convective flow by absorbing and reflecting heat, causing variations in mantle temperature. They also provide a continually changing surface boundary between mantle and crust. Crustal events, such as plates breaking up and slabs subducting, may change the flow of the mantle.

GEOLOGY IN SPACE

If Earth turns out to be a much more active object than we previously thought, how active might other planets be? Looking around our solar system, it appears that Earth is the only planet that is seismically active. Some other planets, such as Mars and Venus, and the Moon, have features that reveal a seismic past: evidence of volcanic activity, for instance. But volcanoes on these planets have been extinct for as many as a billion years.

Some planetary moons also have features that suggest tectonic activity. Io, one of Jupiter's moons, has gas plumes rising high above its surface, and Ganymede, also orbiting Jupiter, has plate-like blocks alternating with trenches on its surface. Both features are reminiscent of places on Earth where shifting plates and the mantle below generate surface activity. More research will reveal whether and how convection is occurring on these moons. For example, if they have liquid interiors, it is more likely that they, too, are lively celestial bodies, with active interiors and surfaces in constant motion. ■

DISCUSSION QUESTIONS

1. How does the work of Alfred Wegener continue to affect what scientists study today?
2. How might the study of tectonic features on other planets contribute to our understanding of Earth?

LESSON 9

INTRODUCING VOLCANOES

WHAT DO YOU THINK THIS VOLCANOLOGIST IS DOING WITH THIS SEISMOGRAPH AT LAKE BUTTE IN YELLOWSTONE NATIONAL PARK?

PHOTO: NPS photo by Jim Peaco

INTRODUCTION

Earthquakes often occur with little warning. Volcanic eruptions, by contrast, can often be forecast well before they happen. Many different signs from the earth tell scientists that a volcano may be about to erupt. Earthquakes, which normally occur before a volcanic eruption, mean that molten rock within the earth is rising and putting pressure on rock. Usually these earthquakes are weak and cannot be detected without the aid of seismographs. An increase in the number of earthquakes may indicate to scientists that a volcano is getting ready to erupt.

Other signs of possible eruption include the presence of steam and ash, which can emerge during small explosions from a volcanic vent. The amount of sulfur in the air over a volcano might also increase as gas is released from the rising molten rock. The top and sides of the volcano may begin to bulge as the molten rock approaches the surface. Volcanologists use special tools to measure the changes that occur in a volcano. By monitoring these changes, scientists can attempt to forecast when the volcano might erupt. The right forecast can save lives and protect property.

What causes volcanoes? How are volcanoes destructive? Do volcanoes have any constructive, or good, effects? In this lesson, you will investigate questions such as these and discuss the relationships among volcanoes and other plate tectonics.

OBJECTIVES FOR THIS LESSON

- Analyze the causes and effects of volcanic eruptions by watching a video.
- Brainstorm what you know and want to learn about volcanoes.
- Analyze scientists' ability to forecast volcanic activity and explore the challenges they face in making such forecasts.
- Identify other plate tectonics related to volcanoes.
- Classify the effects of volcanic eruptions as either destructive (negative) or constructive (positive).

MATERIALS FOR LESSON 9

For you

1 copy of Student Sheet 9.1: Plate Tectonics World Map
Tape

For your group

1 copy of Student Sheet 9.1: Plate Tectonics World Map
1 set of colored markers
1 sheet of newsprint

GETTING STARTED

1. Review what you learned in Lesson 8 about heat in the interior of the earth, then set up two blank pages of your science notebook as directed by your teacher.

2. Tape Student Sheet 9.1: Plate Tectonics World Map to the blank left page of your science notebook, then on the right page, copy the concept map your teacher has drawn.

3. Use what you know about volcanoes to complete the concept map, and on Student Sheet 9.1, plot the locations of where you think volcanoes are found.

4. Discuss your concept map and your world map with your group. Then complete a group concept map on a sheet of newsprint and plot the locations of your volcanoes on the group's copy of Student Sheet 9.1: Plate Tectonics World Map.

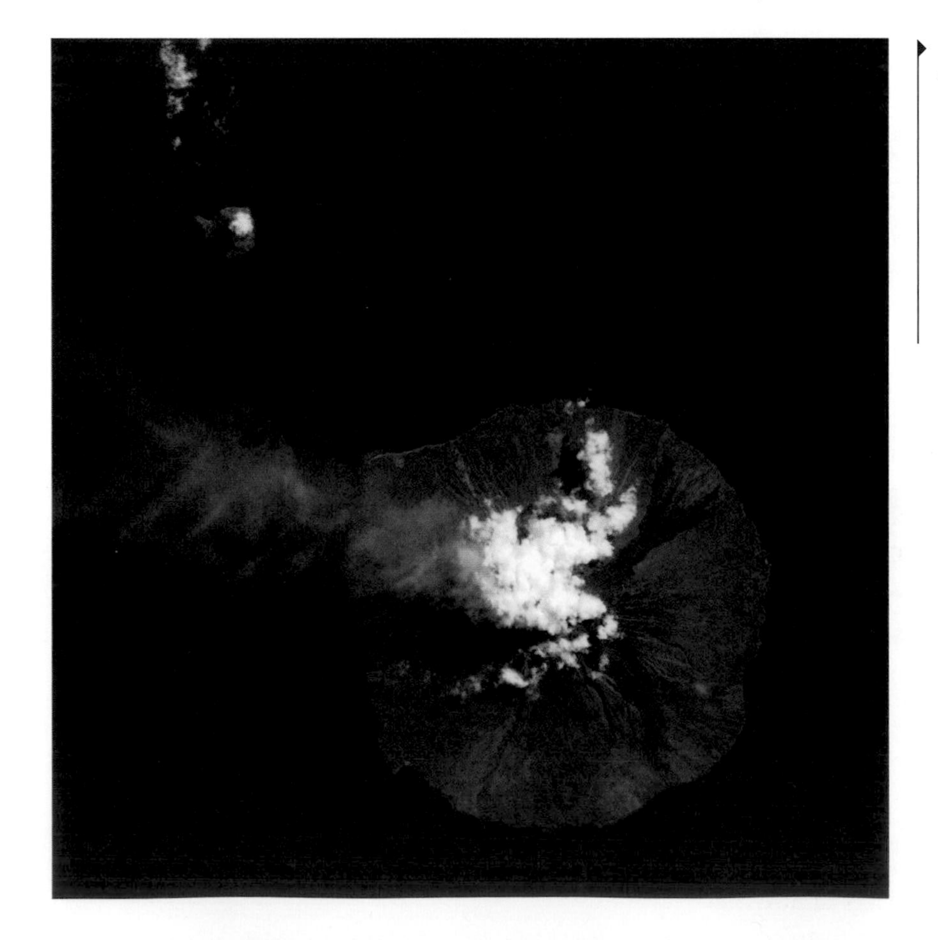

MANAM VOLCANO IS JUST OFF THE COAST OF MAINLAND PAPUA NEW GUINEA. IT STRETCHES 10 KILOMETERS (6 MILES) ACROSS AND IS ONE OF PAPUA NEW GUINEA'S MOST ACTIVE VOLCANOES.

PHOTO: NASA image created by Jesse Allen, using EO-1 ALI data provided courtesy of the NASA EO-1 Team

INQUIRY 9.1

THINKING ABOUT VOLCANOES

PROCEDURE

1. Watch the video about the volcanic eruption of Mt. Pinatubo in the Philippines, called *In the Path of a Killer Volcano.* As you watch, identify the following things and record information about them in your science notebook:
   - A. Two or more instruments or procedures that scientists used to monitor the volcano's activity
   - B. Two or more signs that the volcano was about to erupt
   - C. One or more possible causes of the volcano's eruption
   - D. Two or more effects of the volcano's eruption

   Organize your ideas using a table, list, chart, or other method.

2. Return to your group's concept map. Add to your group concept map any new information you learned by watching the video.

3. Share your group's revised concept map with the class. Your teacher will use your ideas to make a class concept map.

4. Write a definition of "volcano" in your science notebook based on your work and the work with your group and class. Include the evidence that volcanoes reveal that the earth's interior is hot, and how volcanoes shape the earth's crust.

REFLECTING ON WHAT YOU'VE DONE

1. Think about the work of the scientists in the video. Answer the following questions in a class discussion:
   - A. Did the scientists work in a group or alone when observing Mt. Pinatubo? Why do you think they did so?
   - B. How did the scientists monitor the volcano? What were the signs that it would erupt?
   - C. What were some of the risks posed by the volcano's possible eruption?
   - D. What could the scientists do to reduce or eliminate these risks?
   - E. What did the scientists consider when deciding whether to issue an alert?
   - F. What challenges did the scientists face in deciding whether to issue an alert?
   - G. In what way did the scientists communicate their ideas to others?
2. Think about the eruption of Mt. Pinatubo. Describe other plate tectonics that are associated with volcanoes.
3. Watch a second video titled *Geothermal Energy*. How are the effects of volcanoes in this video different from those shown in the first video?
4. Did you gain any additional knowledge about volcanoes from the second video? Your teacher will add those ideas to the class concept map.
5. Work with your class to classify the effects of volcanoes as either constructive or destructive.
6. What questions do you have about volcanoes? Share them with your class. You may find answers to many of these questions as you complete Lessons 11 through 14.

READING SELECTION

EXTENDING YOUR KNOWLEDGE

VOLCANOES: HELP OR HINDRANCE?

Volcanic eruptions can range from violent to mild. All kinds of eruptions have effects that can be both harmful and beneficial to people and the environment.

VOLCANOES CAN BE DESTRUCTIVE

When volcanoes erupt, they often spew molten rock and fragments of rock over the ground and into the air. Fine fragments of rock, called ash, are usually ejected during very violent eruptions. Ash can affect people hundreds of kilometers away from an eruption. In 1980, in Spokane, Washington, it was dark at noon as a result of the ash cloud from the Mt. St. Helens' eruption more than 300 kilometers (186 miles) away. Closer to the mountain, several people died from suffocation by the ash cloud from the initial blast. Volcanic ash can also contaminate water supplies, cause electrical storms, and collapse roofs.

Sometimes a volcano explodes sideways, shooting out ash and large pieces of rock that travel at very high speeds for several kilometers. These explosions can cause death by suffocation and knock down entire forests within seconds. Rivers of molten rock or hot fragments of rock from such eruptions can instantly ignite fires for great distances.

An erupting volcano can also be accompanied by earthquakes, flash floods, rockfalls, and mudflows. Floods occur when rivers are dammed by trees felled during an eruption or by molten rock moving across a river. Mudflows are powerful rivers of mud that form when debris from a volcanic

FOR WEEKS, MT. ST. HELENS SPEWED VOLCANIC ASH OVER THE SURROUNDING LANDSCAPE AND FOR HUNDREDS OF KILOMETERS DOWNWIND TO THE EAST. NOTICEABLE AMOUNTS OF ASH FELL IN 11 STATES. ALTOGETHER, MT. ST. HELENS EXPELLED ENOUGH ASH TO COVER A FOOTBALL FIELD TO A DEPTH OF 240 KILOMETERS (149 MILES).

PHOTO: U.S. Geological Survey/Cascades Volcano Observatory/photo by Peter Lipman

eruption moves into a stream or river. Mudflows can move faster than people can run, and bridges in the path of these flows can be destroyed instantly. One kind of mudflow, called a lahar, happens when rain falls through clouds of ash or when rivers become choked with falling volcanic debris. During the eruption of Mt. St. Helens in 1980, lahars destroyed more than 200 homes, more than 300 kilometers (186 miles) of roads, and 220 kilometers (137 miles) of river channel.

A volcanic eruption can also cause a tsunami. A tsunami is a series of sea waves usually brought on by underwater earthquakes, but volcanoes can cause tsunamis, too. The collapse of an island during a volcanic eruption or the dumping of heavy loads of volcanic debris into the ocean can create massive waves. The 1883 eruption of Krakatoa, a volcanic island in Indonesia between Sumatra and Java, unleashed a tsunami that swept the coasts of Sumatra and Java and drowned more than 36,000 people.

Severe-weather-related events often accompany volcanic activity. These include lightning, thunderstorms, and whirlwinds (including tornadoes). In addition, the heat caused by a volcanic eruption can melt snow and glaciers, which can lead to flooding and landslides. Ash clouds from an erupting volcano can temporarily affect the weather in cities that are hundreds or even thousands of kilometers away. For example, the 1883 eruption of Krakatoa released 20 cubic kilometers (4.8 cubic miles) of volcanic dust into the air. The dust rose so high that it reached the stratosphere. Within 13 days, it had encircled the globe and blocked sunlight from entering the atmosphere. For months, sunsets were strange-colored. Average daily temperatures around the world dropped an estimated 0.5°C (33°F) during 1884. It took five years for all of the volcanic dust to settle to the ground.

IN 1989, THE WAHAULA VISITOR CENTER IN HAWAII WAS ENGULFED BY A HOT LAVA FLOW AND BURST INTO FLAMES. ALL ATTEMPTS TO SAVE THE CENTER WERE USELESS.

PHOTO: U.S. Geological Survey/photo by J.D. Griggs

READING SELECTION

EXTENDING YOUR KNOWLEDGE

AFTER THE ERUPTION OF MT. ST. HELENS, THIS HOME WAS DAMAGED BY VOLCANIC MUDFLOW ALONG THE SOUTH FORK TOUTLE RIVER IN WASHINGTON STATE.

PHOTO: U.S. Geological Survey/Cascades Volcano Observatory/photo by Lyn Topinka

In 1815, a different Indonesian volcano, Tambora, erupted even more powerfully. It blasted about 150 cubic kilometers (36 cubic miles) of volcanic debris high into the atmosphere. The dust blocked so much sunlight that crops failed to grow around the world, and 1816 became known as "the year without a summer." Again, it took several years before the effects of this eruption passed.

VOLCANOES CAN BE CONSTRUCTIVE

Not all the materials that come out of volcanoes are harmful. Many volcanic areas have permanent hot springs that are beautiful to look at and provide recreation for residents and tourists. In addition, people can tap the geothermal energy of hot springs to heat their homes directly or to produce electricity. Icelanders, for example, use geothermal energy to heat their homes, buildings, and swimming pools. Iceland has a very short growing season, but greenhouses heated by geothermal energy provide Icelanders with vegetables, tropical fruit, and flowers year-round. Some people living in Arctic regions also heat their homes and greenhouses with water from hot springs. The hot water flows through pipes in their houses, warming the air. Geothermal steam is used to generate electricity in places such as Italy, New Zealand, the United States, Mexico, Japan, and Russia.

Volcanoes provide a wealth of natural products. Basalt, which forms from cooled lava and makes up much of the seafloor, is a raw material for cleaning agents, and it has many chemical and industrial uses. Volcanic ash enriches the soil with mineral nutrients. Minerals in molten rock are a major world source of nickel, chromium, platinum, and several other important elements. Obsidian, or "volcanic glass," is an ideal material for fine stonework because it breaks with a typical curved fracture when struck with a sharp blow. Beautiful arrowheads of obsidian have been found in Ohio

MOST GEYSERS ARE HOT SPRINGS THAT ERUPT FOUNTAINS OF SCALDING WATER AND STEAM.

PHOTO: NPS photo by Frank Balthis

THIS OLD ENGRAVING SHOWS THE 1866 ERUPTION OF NEA KAMENI, SANTORINI, IN GREECE. A GIANT VOLCANIC EXPLOSION CAUSED THE SUDDEN SINKING OF THE ISLAND'S CENTER BENEATH.

PHOTO: P. Hedervari, National Geophysical Data Center/NOAA

from the Hopewell culture, which flourished 1500 to 2300 years ago.

Volcanoes also create beautiful landscapes. Without volcanic activity, there would be none of the spectacular fissures that dot the Hawaiian landscape or the majestic peaks of the Cascade Range, such as Mt. Rainier.

Most people think of catastrophic events as violent natural hazards that create human and environmental risks. But as we have just seen, there is another side of the story. Catastrophic events can also be constructive forces on the earth. Volcanoes affect the composition of our oceans and atmosphere. Floods create sandy beaches along riverbanks. And earthquakes, as well as volcanoes, create and shape the mountains and islands that people enjoy. ■

DISCUSSION QUESTIONS

1. The primary effect of a volcanic eruption is the spewing of lava, rocks, and/or ash. What are some secondary effects of a volcanic eruption?
2. What are two ways in which volcanoes are seen as beneficial?

READING SELECTION

EXTENDING YOUR KNOWLEDGE

EARTH'S WATERWORKS

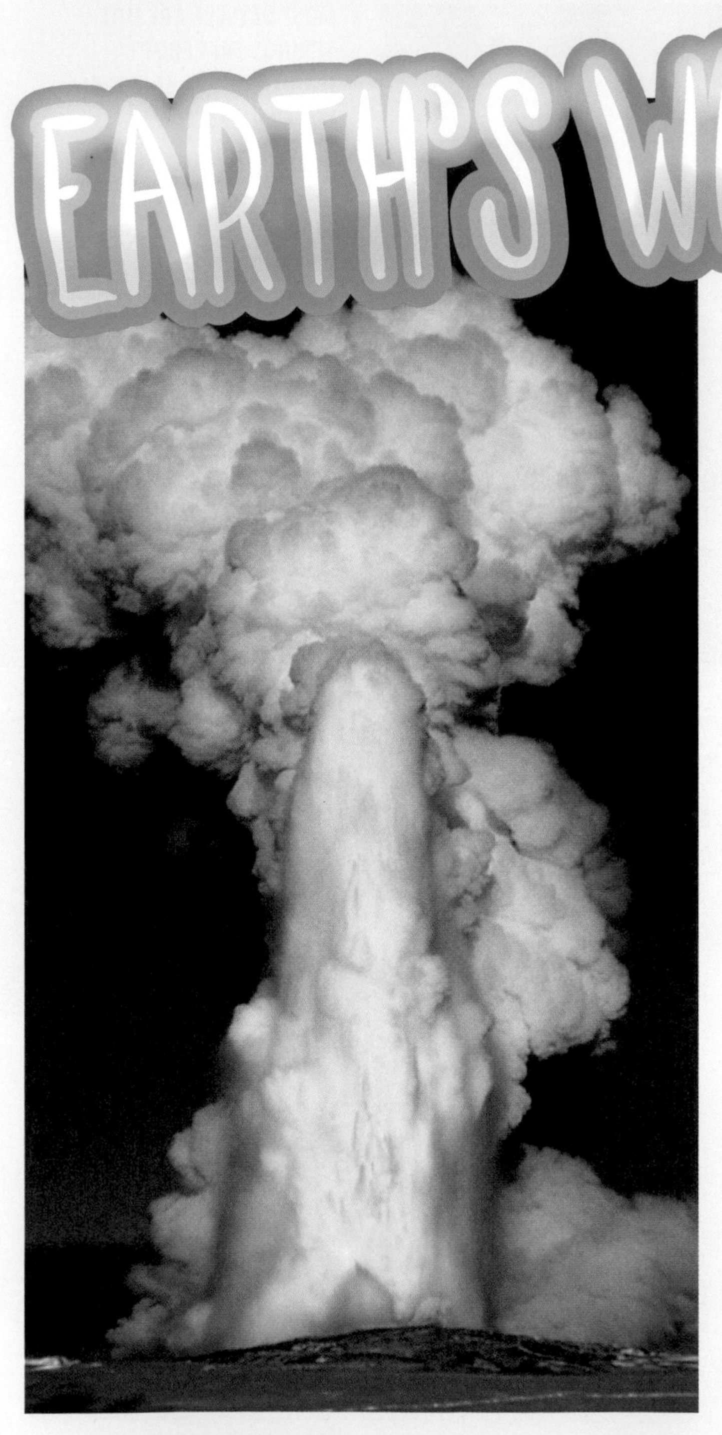

OLD FAITHFUL IS THE MOST FAMOUS GEYSER IN YELLOWSTONE NATIONAL PARK.

PHOTO: NPS Photo

In the 1800s, fur trappers who ventured into the Rocky Mountains came back and told of "the place where hell bubbles up." No one believed them. One of the trappers, a Virginian named Jim Bridger, told people about finding a column of water as thick as his body that spouted 18 meters (about 60 feet) in the air. People called him a liar. But explorers later confirmed the trappers' stories. They had found the roaring geysers (springs that spout hot water and steam) and craters of boiling mud in the area that is now Yellowstone National Park. Today, several million people every year come to view these wonders at Yellowstone.

Yellowstone Park, located in Wyoming and Montana, is the hottest, most active geyser area in the world. It contains more than 500 geysers, or nearly three-fourths of all the world's geysers. In all, Yellowstone has 10,000 geothermal features ("geo" means "earth" and "thermal" means "heat"). Besides geysers, Yellowstone's geothermal features include hot springs and bubbling mud pots.

Why does Yellowstone have so many hot springs? Most of Yellowstone sits inside an ancient caldera. The volcano's last major eruption, which created the caldera, happened 600,000 years ago. Smaller lava flows from the volcano gradually filled up most of the caldera. Rock under the earth's surface can stay hot for thousands of years. Heat from molten rock a few kilometers below the surface heats the groundwater. The groundwater is held in a porous type of rock, and the heated water travels upward until it bursts through the earth's surface like a fountain.

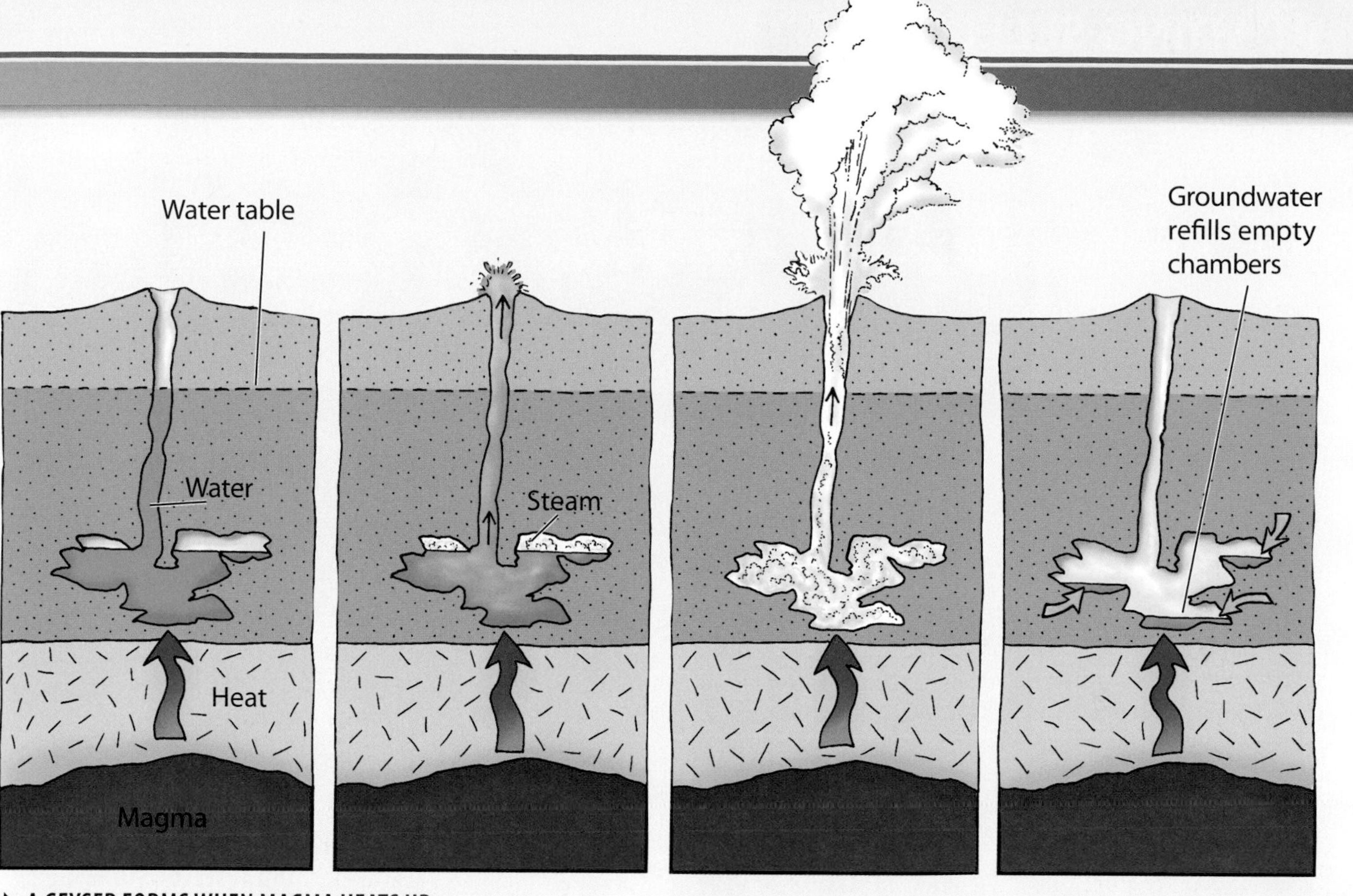

A GEYSER FORMS WHEN MAGMA HEATS UP GROUNDWATER THAT IS UNDER PRESSURE.

Old Faithful, Yellowstone's most famous geyser, got its name because it normally erupts on a regular basis—on average, every 79 minutes. The eruptions are so regular because the water supply and the structure of the rock remain fairly constant over time. Yet Old Faithful isn't completely predictable. The time between eruptions actually varies between 45 and 105 minutes, depending on the amount of super-hot water left in the spongy rock when the geyser runs out of steam.

Recently, scientists lowered a video camera and other instruments into the vent of Old Faithful. They found that for the first 20 or 30 seconds of each eruption (which lasts for several minutes), steam and boiling water rocket through the narrowest underground cracks at the speed of sound!

WHAT MAKES A GEYSER GO?

Rainwater trickles through cracks into porous rock, where it collects like water in a sponge. Heat from magma a few kilometers beneath the earth's surface rises and heats the water in the porous rock. The porous rock layer is like a pressure cooker: it has lots of heat from the magma and lots of pressure from the weight of water and rock above it. The water in the porous rock can reach temperatures of 310°C (590°F) without boiling because of the tremendous pressure.

This super-hot water rises into pockets of groundwater that are also under pressure. Steam forms, more pressure builds, and bubbles rise. Steam keeps building until a spout of hot water and vapor explodes to the surface and shoots high into the air. More super-hot water then bursts into steam and blasts more groundwater out of the earth, erupting sometimes for up to several hours.

READING SELECTION

EXTENDING YOUR KNOWLEDGE

HOT SPRINGS ARE ONE OF THE GEOTHERMAL FEATURES FOUND AT YELLOWSTONE NATIONAL PARK.

PHOTO: NPS photo by J.R. Douglass

WHEN THE WATER FROM CASTLE GEYSER IN YELLOWSTONE NATIONAL PARK EVAPORATES, A MINERAL CALLED SILICA IS LEFT BEHIND. THIS MOUND OF SILICA IS CALLED A GEYSERITE.

PHOTO: NPS photo by George Marle

If the super-hot water mixes with cool groundwater that is not under pressure, it rises to the surface as a hot spring. When hot springs become choked with pieces of weathered rock (sediment) that break off from the surrounding rock, bubbling mud pots are the result.

If the super-hot water rises to the surface with no resistance, it begins to boil and erupts at the surface as steam. This thermal feature, called a fumarole, is like a geyser, except that it is mostly steam.

WHEN GEYSERS LOSE THEIR STEAM

Some old geysers lose their steam. The super-hot water carries minerals that, through time, accumulate on the walls of the underground channels and cracks. Like arteries, the cracks become clogged, and the steam and water can no longer escape.

For one Yellowstone geyser, named Porkchop, the pressure was too much. It spouted water and steam for years. Then one day it blew rocks the size of TV sets into the air and stopped gushing for good. ■

EARTHQUAKE PREDICTOR?

In 1990, a scientist from the Carnegie Institution in Washington, DC, worked with a person who owned property near a geyser in California to study patterns in the geyser's eruption cycle. The geyser didn't always erupt on schedule. For 15 years, the property owner had collected data on the geyser's eruptions. The scientist compared the data with records of thousands of earthquakes in California and found that major changes in the geyser's activity coincided with three large earthquakes that occurred within 248 kilometers (154 miles) of the geyser. In all cases, changes in the geyser happened one to three days before the earthquake. The scientist hypothesized that underground movements that caused the earthquakes may also have affected the geyser's water supply.

Earthquakes often hit the Yellowstone National Park area, shaking and moving the geysers' "plumbing system" and choking off the water supply. An earthquake in 1995 moved heat and water away from Steamboat Geyser and redirected it to Monarch Geyser, which had been dormant for 81 years. Suddenly, Monarch began blowing off steam. Could a geyser be an earthquake detector? It's possible, but more study needs to be done on this subject.

DISCUSSION QUESTIONS

1. Why do geysers erupt? Why don't they remain as pools of hot underground water?
2. Imagine you are a scientist tasked with studying how geysers might serve as earthquake detectors. What sort of data could you gather?

LESSON 10

EXPLORATION ACTIVITY: EXPLORING MITIGATION OF EARTHQUAKES AND VOLCANIC ERUPTIONS

INTRODUCTION

On December 26, 2004, the fourth largest earthquake since 1900 rocked the region around Sumatra and triggered a tsunami that reached 14 countries. The landscape of Banda Aceh was permanently altered. The event killed 227,898 people, and 1.7 million people were displaced. The earthquake registered 9.1 on the Richter scale, and people in Banda Aceh experienced a Mercalli rating of IX. As tragic as this event was, valuable lessons were learned and heeded. Was this disaster preventable? What technologies and designs have we developed as a result of tragedies like this one?

AN AERIAL VIEW OF THE DESTRUCTION IN BANDA ACEH, INDONESIA AFTER A 2004 TSUNAMI.

PHOTO: Patrick M. Bonafede/U.S. Agency for International Development (USAID)

In this activity, you will research an historic earthquake or volcanic eruption. Then you will relate the event to a specific technology, building design, or monitoring system that has been used to prevent the loss of lives in future earthquakes or volcanic eruptions. You will work with your group to give a historical perspective about the event, detail the event's impact on people's lives and their surroundings, and describe the technology used to monitor and predict these disasters. You will develop a detailed display summarizing your research and present it to the class.

OBJECTIVES FOR THIS LESSON

- Read about the Sumatran tsunami of December 26, 2004.
- Define mitigation and link a specific mitigation to a specific seismic or volcanic event.
- Review the Exploration Activity guidelines.
- Select a topic of study for the Exploration Activity.

MATERIALS FOR LESSON 10

For you

1 copy of Student Sheet 10.1a: Exploration Activity Timeline

For your group

1 copy of Inquiry Master 10.1a: Exploration Activity Scoring Rubrics

1 copy of Student Sheet 10.1b: Exploration Activity Research Questions

1 copy of Student Sheet 10.1c: Exploration Activity Group Roles

GETTING STARTED

1 Read "Reading the Signs: Detecting Seismic Events" on pages 134–139. Discuss your ideas about what happened to Tilly Smith at the beach.

A. How would you define a "tsunami"?

B. Many lives were lost. Could anything have been done to prevent the scale of the tragedy?

C. When talking about natural disasters, the word "mitigation" is used to mean actions to reduce the loss of life and property. What are some possible mitigations for various catastrophic events? Star the ones specifically related to earthquakes and volcanoes.

AFTER A SERIES OF EARTHQUAKES IN 2006, ENGINEERS INSPECT A TRANSFORMER IN HAWAII TO MAKE SURE THAT IT REMAINS OPERATIONAL AT A SITE THAT TRANSMITS EMERGENCY MESSAGES.

PHOTO: FEMA/Adam Dubrowa

PART 1
PLANNING YOUR RESEARCH PROCEDURE

1. Your teacher will tell you about the Exploration Activity, a group research project that focuses on reducing the risks associated with severe seismic or volcanic events. Discuss these points with your teacher:

   A. You will work in groups and begin the Exploration Activity today.

   B. Each group will select one historic earthquake or volcanic eruption. You will research the event using the Exploration Activity Guidelines. Based on your research, you will report on the facts of the event, the impact the event had on people, and the effectiveness of the programs designed to mitigate the event.

   C. You will conduct most of your research outside of class.

   D. Each group will present its Exploration Activity to the class on a date determined by your teacher.

2. Read the Exploration Activity Guidelines on pages 128-129. Then review them with your teacher.

3. Your teacher will give you a copy of Student Sheet 10.1a: Exploration Activity Timeline. Each member of your group will be responsible for completing the activity. Discuss the due dates with your teacher. Make certain you record them on your timeline.

4. Spend a few minutes with your group deciding on an event to research. Examples are listed in the Exploration Activity Guidelines.

5. Discuss with your teacher how your work will be assessed for the Exploration Activity. Review Inquiry Master 10.1a: Exploration Activity Scoring Rubrics.

6. For your group, get one copy of Student Sheet 10.1b: Exploration Activity Research Questions and Student Sheet 10.1c: Exploration Activity Group Roles. Each group is responsible for turning in these sheets on the assigned due dates. Record the due dates at the top of each sheet.

7. Discuss with your group Student Sheet 10.1b. Discuss ways you can find out the information to answer the questions on the sheet.

Exploration Activity Part 1 continued

EXPLORATION ACTIVITY GUIDELINES

You will first select an earthquake or volcanic eruption. In your research of an earthquake, you will identify and describe the science behind it: plate motions, type of fault involved, Richter and Mercalli scales, etc. In your research of a volcanic eruption, you will describe the science behind it: where it is located and the plate motions, the type of volcano, etc. You will research the impact of your selected event on people and on the surrounding area. You will also explain the technology or preparedness program for the affected area and evaluate how well it worked. Finally, you will present your findings to the class.

PROJECT IDEAS

You can select from a variety of categories: United States, world, deadliest, largest, etc. These categories can be a starting place to investigate one of the earthquakes or volcanic eruptions listed below, or to find another earthquake or volcanic eruption. You will need to explain and evaluate a technology or preparedness program that relates to your event. Some programs and technologies to mitigate these disasters are also listed below. You may be interested in researching an event that links to one of these specific programs.

EARTHQUAKES

1700	Cascadia Subduction Zone, Washington	1989	Loma Prieta, California
1811	New Madrid Region, Missouri	1991	Costa Rica
1906	San Francisco, California	1993	India
1920	Haiyun, Ningxia, China	1995	Kobe, Japan
1923	Kanto, Japan	2004	Sumatra
1949	Puget Sound, Washington	2005	Pakistan
1964	Prince William Sound, Alaska	2006	Indonesia
1976	Tang Shan, China	2008	Sichuan, China
1976	Guatemala	2010	Haiti
1985	Michoacan	2011	Japan

VOLCANIC ERUPTIONS

1640 BC	Island of Thira, Aegean Sea	**1902**	Montagne Pelée, French West Indies
79 AD	Mt. Vesuvius, Italy	**1951**	Lamington, Papua New Guinea
1783	Laki, Iceland	**1985**	Nevado Del Ruiz, Colombia
1815	Mt. Tambora, Indonesia	**1991**	Mt. Pinatubo, The Philippines
1883	Krakatoa, Indonesia	**1994**	Mt. Rabaul, Papua New Guinea

TECHNOLOGY/EDUCATION/PREPAREDNESS

- Steel reinforced concrete
- Tsunami early-warning network buoys
- Earthquake-proofing: bridges, railroads, houses
- Search and rescue devices
- Search and rescue procedures
- Emergency management systems
- Base isolators
- Energy-dissipation devices
- Seismic dampers
- Active-control devices
- What to do in the event of earthquakes, volcanic eruptions, or other resulting phenomena, such as tsunamis and landslides
- Search and rescue techniques and coordination
- Earthquake prediction/fault monitoring technology
- Volcano prediction/monitoring technology
- Retrofitting techniques for existing structures

PART 2
CONDUCTING YOUR RESEARCH
PROCEDURE

1. You should conduct your research using a variety of resources, including those found in the classroom, your school library and other libraries, newspapers, or on the Internet. You may also conduct personal interviews.

2. Your teacher will tell you how many Internet sources, books, magazines, and newspaper articles you need to use in your research.

3. Your teacher will also tell you how many images you need to include in your final presentation and report. An image can come from an instructional technology tool (such as a CD-ROM, website, or video) or from a magazine or newspaper article.

4. Here are some specific questions your group should answer while conducting your research. Use this information to write your report.

A. When, where, and how did the earthquake or volcanic eruption occur? What caused it? Apply what you've learned throughout the unit when discussing the event you've selected:

   1. Describe the surface movement, magnitude, and intensity of the quake or eruption.
   2. Name the plate(s) and describe the plate motions involved.
   3. Name and describe the type of fault.
   4. Give the name of the fault if you can find it.

B. What were some of the risks of this event?

   1. How frequent is the activity along this fault?
   2. Describe what happened during the earthquake or volcanic eruption to affect human lives. Include details of what structures and systems failed.
   3. What physical damage and environmental effects resulted from the event?

C. How did people respond to the event?

   1. Name and describe the agencies, federal and international, involved in the rescue and recovery efforts.
   2. Were there systems and/or structures in place prior to the earthquake or volcanic eruption that may have lessened the impact of the event, or may have assisted with the recovery of the region?

D. What have people learned from the event to reduce future risks? For example:

1. How do people forecast future events like this one?
2. How do people reduce or prevent future risks associated with events like this one? Consider a specific manmade mitigation or environmental/development protections.
3. How have people used technology to reduce the risks associated with events like this one?

5 Your teacher may give you time during class over the next couple of weeks to conduct some of your research. You will, however, be required to conduct most of the research outside of class.

PART 3

PRESENTING WHAT YOU HAVE LEARNED

PROCEDURE

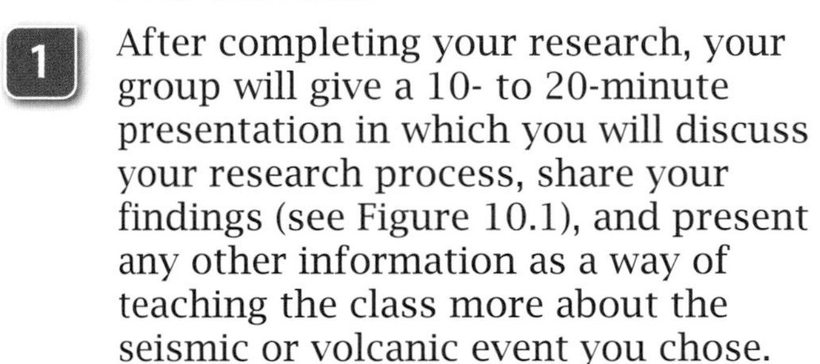

1 After completing your research, your group will give a 10- to 20-minute presentation in which you will discuss your research process, share your findings (see Figure 10.1), and present any other information as a way of teaching the class more about the seismic or volcanic event you chose.

A STUDENT DISPLAYS A MODEL OF A VOLCANOLOGIST'S TILTMETER, WHICH MEASURES CHANGES IN THE SLOPE OF THE VOLCANO. A CHANGE IN SLOPE MAY BE A SIGN THAT A VOLCANO IS ABOUT TO ERUPT.
FIGURE **10.1**

PHOTO: © Terry G. McCrea/Smithsonian Institution

Exploration Activity Part 3 continued

2. Each member of your group should play a role in this presentation.

3. Each group member is expected to turn in a written summary of his or her research.

4. Your teacher will tell you the minimum number of pages for your report or will give you guidelines for producing an electronic presentation.

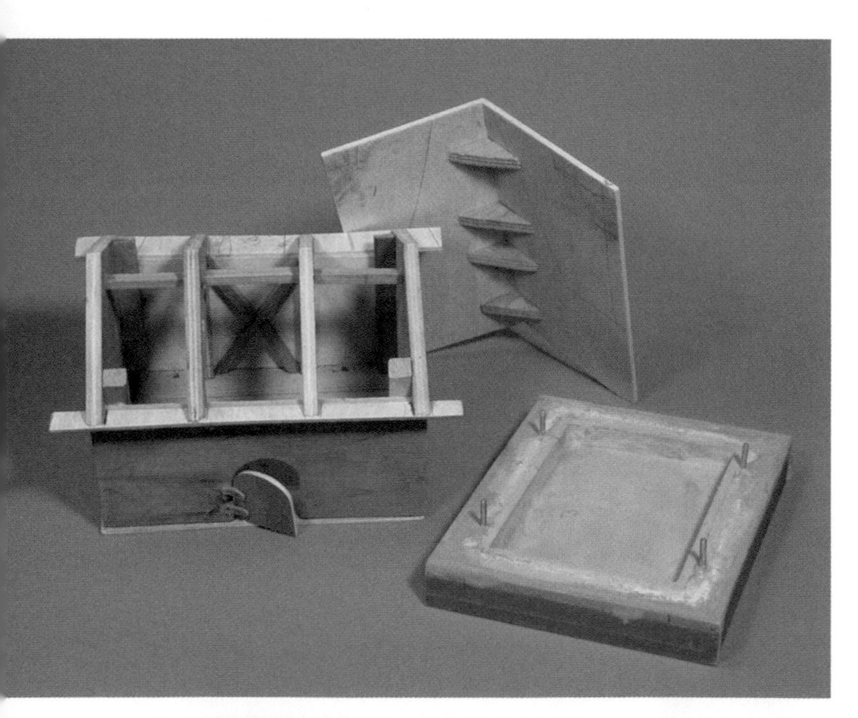

A STUDENT MODEL OF A HOUSE THAT WAS REINFORCED TO MAKE IT EARTHQUAKE RESISTANT. CROSSBARS ARE INSIDE THE WALLS TO MAKE THEM STRONGER. THE HOUSE IS FIRMLY SECURED TO THE FOUNDATION, WHICH IS CONCRETE.
FIGURE **10.2**

PHOTO: © Terry G. McCrea/Smithsonian Institution

5. Here are some ideas that may help in developing your own presentation:
   - Hold a town meeting to debate what was learned from the seismic event. Each member of your group can take on a different role, such as town mayor, town citizen, insurance agent, building inspector, and business owner. (See the example on Student Sheet 10.1a.)
   - Create a model of a building that would withstand the seismic event.
   - Design a newscast, complete with a discussion of the seismic activity of the area, advice on managing trauma in the aftermath, a report on the latest advancements in seismic detection technology, educational segments reporting on the environmental effects and the steps we can take to better utilize the environment for protection, as well as how to prepare for future events and rebuild after an event.
   - Hold a panel discussion in which you role-play scientists, citizens, building engineers, and representatives from corporations who debate the trade-offs and solutions for risk reduction. For example, do you require the retrofitting of buildings in your area, or just require that new construction projects be earthquake-resistant?

- Develop literature to promote awareness of the event. For example, write newspaper articles reporting on the event, do a television news report, or offer a radio editorial describing social and personal impacts of the event.
- Design a kit for citizens' emergency preparedness.
- Research the role of humanitarian groups in mitigating effects of volcanoes and earthquakes.
- Develop a database listing similar events around the world and a map or other visual to display those data. For example, you may want to visit the U.S. Geological Survey's earthquake website for a database of earthquakes that occurred this year. Discuss the patterns of these events. How are they alike? What did we learn from them?
- Propose a design (and possible physical model) for a product or tool that would either help reduce the risks associated with an earthquake or volcanic eruption, or that will assist with the rescue and recovery efforts.
- Design and build a museum exhibit to display information about your event.
- Show images of equipment currently used by scientists to detect severe seismic events.

6. Be sure to practice your presentation with your group members before giving it to the class.

7. Your teacher will tell you the day on which you will give your presentation.

READING SELECTION

EXTENDING YOUR KNOWLEDGE

READING THE SIGNS: DETECTING SEISMIC EVENTS

The next time your mind is wandering during class, consider what you may be missing. In December of 2004, a ten-year-old British girl was able to save over a hundred people, including her family, thanks to what she'd learned from a video she'd seen in her geography class. The video showed a tsunami approaching shore; a tsunami is a series of enormous, fast-moving ocean waves generated by an undersea earthquake or volcanic eruption. They can be tremendously destructive. Although tsunami waves may not be tall waves in the open ocean, as they knock up against the rising seabed of the shore, they slow down, grow tens or hundreds of feet tall, and begin to pile up behind each other. As they gather for the crash on the shore, they may suck water back into the ocean, leaving people on the beach wondering why the tide's suddenly gone out so far.

Some time after Tilly Smith saw the tsunami video in class, her family took a winter seaside vacation in Thailand, on the Indian Ocean. While playing on the beach, Tilly noticed that the ocean was rising and becoming bubbly and frothy. The scene that unfolded before her eyes looked alarmingly like what she had seen in the video. In a panic, she screamed for her family to get off the beach. Luckily, her parents took her fright seriously, warned other vacationers and guards, and ran up the beach to their hotel. Other beachgoers followed. While they were sheltered on the third floor of the hotel, the beach was pummeled by the tsunami. It was one of the few beaches hit that day on which no one was killed.

The giant waves were caused by a powerful earthquake (magnitude-9) that had occurred in Sumatra, an island in another part of the Indian Ocean, earlier that day. Although Tilly's ability to read the signs of a tsunami got everyone off her hotel beach to safety, nearly 228,000 people in eleven countries were killed by the tidal waves from the Indian Ocean during the catastrophic event.

If a ten-year-old was able to see a tsunami coming and clear a beach, why couldn't more people have been saved that day? The answer lies partially in how difficult it can be to spot major catastrophes in time to warn authorities and have them get people to safety. Tsunamis

PEOPLE WERE STILL DIGGING THEIR BELONGINGS OUT OF RUINED HOUSES SEVERAL WEEKS AFTER THE 2004 TSUNAMI IN THE INDIAN OCEAN.

PHOTO: U.S. Navy photo by Photographer's Mate Airman Jordon R. Beesley

are huge volumes of water pushed into motion by a catastrophic event undersea, but most of that rolling mass lies under the ocean's surface. In the open ocean, tsunami waves may look like normal waves, only a foot or so tall. The bulk of the enormous wave lies underneath. It's only when the mass is forced up and out of the ocean by a rising shoreline that the size and force of the wave is apparent.

READING SELECTION

EXTENDING YOUR KNOWLEDGE

There are signs, though, that such enormous waves are on the move under the ocean's surface. Water pressure may change suddenly undersea as the wave passes by. Being able to read this tsunami signal can help in setting up effective early warning systems. Many countries participate in a system designed to detect changes in ocean water pressure that signal an impending tsunami. In this four-part system, seafloor sensors constantly monitor water pressure and send signals to the surface buoys, which then relay the data to satellites, which send the data to tsunami warning stations.

Through its DART (Deep-ocean Assessment and Reporting of Tsunamis) program, the U.S. National Oceanic and Atmospheric Administration manages a set of sensors around the "ring of fire" in the Pacific, where plates collide. Its tsunami warning stations are staffed 24 hours a day, looking for signs of trouble. When a 2010 earthquake in Chile spawned tsunamis in the Pacific Ocean, DART buoys were able to detect them and warn people in Hawaii ten hours in advance. A loud siren blared from the tsunami warning station, giving people time to evacuate coastal areas for higher ground. While the tidal waves that reached Hawaii during that event did not amount to much, the residents were better safe than sorry.

BUOY THAT FORMS PART OF NOAA'S DART II SYSTEM

PHOTO: NOAA

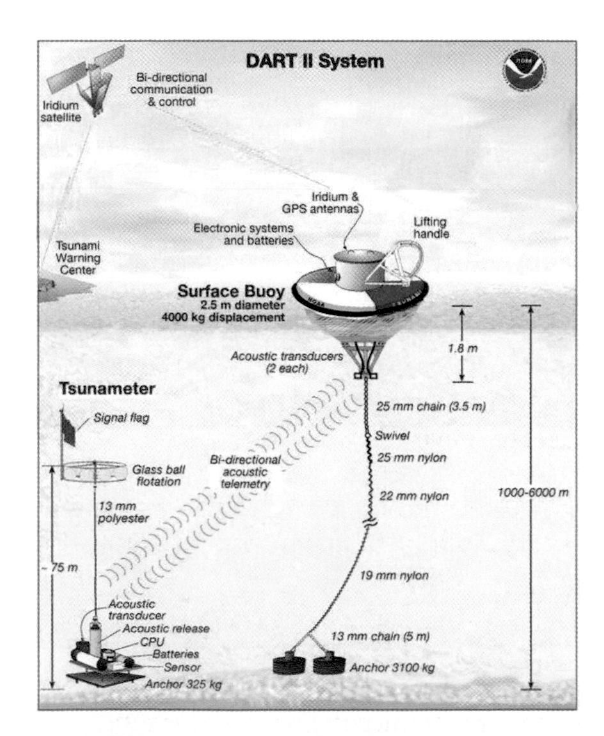

DIAGRAM SHOWING HOW THE DART SYSTEM TRANSMITS SIGNALS FROM THE OCEAN BOTTOM TO A TSUNAMI WARNING STATION

PHOTO: NOAA

AN ELECTRONIC TILTMETER ON THE POST IN THE FOREGROUND MONITORS CHANGES IN SLOPE ON MT. ST. HELENS IN WASHINGTON STATE.

PHOTO: U.S. Geological Survey

Such systems cost money to run and require trained, well-educated staff. In the absence of popular and governmental support, they may not be built. In 2004, there was no DART system in the Indian Ocean; the 2004 tsunamis, and the staggering loss of life in the coastal areas affected, gave the world a graphic demonstration of warning systems' importance. Governments around the region worked with the United Nations and other countries to install an Indian Ocean DART.

While early warning systems for tsunamis can be quite effective, what about warnings for other types of seismic events? There are a number of signs that can be monitored to determine whether a volcanic eruption is coming. For example, gases, particularly sulfur dioxide, are vented from volcanoes as magma shifts and rises to the surface. An eruption is often preceded by a marked change in the amount and composition of gases from the volcanic cinder cone. The build-up of magma below the surface also causes a volcano to swell, a change that can be measured with a tiltmeter (an instrument that detects slight changes in slope). Movements of magma can cause shaking, or tremors, of parts of the volcano, which can be measured with the same sort of seismic equipment used for earthquakes.

Together, these signs—gases, tremors, and swelling—reveal that a volcano is becoming active. However, it's still not possible to predict with accuracy exactly when, or indeed whether, an eruption will occur; nor can we tell how severe the eruption might be. (Often, magma cools on the surface without erupting.) What volcanologists can do is tell us when an eruption is probable. The timing of an eruption depends

READING SELECTION

EXTENDING YOUR KNOWLEDGE

THE GALERAS VOLCANO (GREEN WITH REDDISH CONE) HAS ERUPTED MORE THAN 20 TIMES SINCE THE 1500S. IT IS ONE OF 15 VOLCANOES WORLDWIDE THAT ARE TARGETED FOR MONITORING BECAUSE THEY POSE HIGH RISK TO PEOPLE. NOTE THE CITY OF PASTO (ORANGE AT BOTTOM OF IMAGE) IS JUST 8 KILOMETERS (5 MILES) FROM THE VOLCANO.

PHOTO: NASA Jet Propulsion Laboratory

on many conditions below the surface, such as temperature, structure of the rock overlaying the magma, and amount of pressure exerted by the magma.

Consider the unfortunate events of 1993, when seven volcanologists were studying the active Galeras Volcano in Colombia, South America. They were collaborating to learn more about the early warning signs of eruption. Because the recent tremors and venting of gases had abated, the scientists thought it was safe to climb into the cone to take measurements. They were wrong. A massive explosion killed all but one of the scientists, Stanley Williams, who had to crawl out with broken legs and burned skin. It is thought that the Galeras Volcano had appeared to be "asleep" because magma had sealed up all the surface cracks, temporarily blocking the exit of gases and stilling the tremors. As magma continued to rise inside, the extreme pressure brought on by this situation literally burst the volcano open.

Developing early warning systems for earthquakes has also proven to be difficult. For decades, scientists have been developing technology to monitor the seismic changes that accompany earthquakes. Earthquake monitoring stations now use specialized instruments to detect P-waves, the fast-moving waves that precede the arrival of the more damaging, slower waves. Unfortunately, because the more destructive waves follow right on the heels of the P-waves, at best we can detect an earthquake less than a minute before its destructive impact. This falls short of what we need for an early warning system. Would you be able to put down what you are doing and evacuate to a safe place with less than a minute of warning? The good news is that in recent years, scientists have found that by positioning highly sensitive instruments as deep as a kilometer underground, they can measure subtle changes in stresses on rocks that might signal a quake.

With continued seismological research, we should be able to develop earlier warning systems for earthquakes as well as improve systems for volcanoes and tsunamis. Scientists continue to work on understanding the sometimes hard-to-read signs that a major seismic event is going to occur. As a professor at Arizona State University, the recovered Stanley Williams still climbs into volcano cones collecting data. The work that he and other scientists conduct may save a lot of lives by making it possible to forecast catastrophic events in a timely, accurate way. ■

DISCUSSION QUESTIONS

1. The video Tilly watched in school helped her spot danger and warn her parents of the tsunami. What else could have helped people get off the beach to safety?

2. Scientific research is often expensive, and we cannot fund all the research we might like to fund. Suppose a friend said, "We're really far from being able to predict earthquakes. We should spend money on science that has a better chance of success." Do you agree or disagree?

LESSON 11

VOLCANOES CHANGE THE LANDSCAPE

PILLOW LAVA ROCKS THAT FORMED OFF THE COAST OF HAWAII

PHOTO: OAR/National Undersea Research Program (NURP)

INTRODUCTION

The hot molten rock that lies deep within the earth rises through fractures in the earth's crust. Sometimes it stays below the earth's surface, where it cools slowly and forms new rock. At other times, it spews out onto the land or the ocean floor. This is how volcanic mountains or islands form. If the molten rock flows into the ocean or emerges under the sea (such as along mid-ocean ridges), it cools as soon as it hits the water. In recent years, scientists have used remote-controlled cameras to observe how red-hot liquid rock emerging from the ocean floor cools and turns into a solid mound of lava in only a few seconds. Scientists call the resulting underwater balloon-shaped mounds of lava "pillows."

In this lesson, you will investigate land formation, which is one of the constructive effects of volcanoes. In the first inquiry, you will use a substance called Model Magma™ to simulate how rising molten rock changes the shape of the land above it. In the second inquiry, you will use melted wax to model how cooled molten rock creates new landforms—both on land and under water. Through these investigations, you will see that volcanoes can change the landscape of the earth.

OBJECTIVES FOR THIS LESSON

- Brainstorm what you know and want to learn about magma and lava.
- Model the movement of molten rock through fractures in the lithosphere, over the earth's surface, and under water.
- Devise working definitions for the terms "magma" and "lava."
- Identify in photographs landforms created by molten rock.

MATERIALS FOR LESSON 11

For you

- 1 pair of safety goggles

For your group

- 1 plastic box with lid
  - 1 empty, wide-mouthed, clear plastic container
  - 1 block of wax
  - 2 hand lenses
  - 1 index card
  - 1 sheet of waxed paper
  - 1 plastic spoon
- 2 wide-mouthed, clear plastic containers of soil (with hole in base of each container)
- 1 wide-mouthed, clear plastic container of room-temperature Model Magma™
- 2 containers of melted wax, with lids
- 1 wide-mouth, clear plastic container of heated Model Magma
- 1 beaker of very cold water
- 2 pieces of transparent tape

GETTING STARTED

1. Think back to the video *Geothermal Energy* that you watched in Lesson 9. What were some of the beneficial effects of volcanoes and molten rock?

2. Brainstorm what you know and want to learn about magma and lava.

LAVA ERUPTS FROM A FISSURE BETWEEN PU`U `O`O CRATER AND NĀPAU CRATER IN HAWAII.

PHOTO: U.S. Geological Survey

INVESTIGATING MAGMA AND NEW LANDFORMS

PROCEDURE

1. Look at the sample of Model Magma™. You will use this substance to model how molten rock under the earth's surface affects the land above it. Describe its properties. Consider the following:
   - A. Is it a solid or a liquid? Give two reasons to support your answer.
   - B. How does it differ from other substances you have observed?
   - C. How might you use this substance to model the behavior of molten rock beneath the earth's surface?

2. After reviewing the Procedure and Safety Tips, decide how you might organize your observations in a table. Design the table in your science notebook. Remember to include space for the following:
   - A. Qualitative observations and labeled drawings
   - B. Observations of the hot and cold Model Magma
   - C. Observations of Model Magma before being pushed through the soil
   - D. Observations of Model Magma after being pushed through the soil

3. Pick up your group's box of materials, two containers of soil, and one container of room-temperature Model Magma.

4. Observe the Model Magma in your group's box of materials. Record your observations of the Model Magma. Make a prediction about what might happen to the substance when it is heated. Discuss this with your group. Return the Model Magma to its container. Make certain the soil is firmly pressed into each container. If it is not, pack it down with the spoon. You will use this soil to model land on the earth.

SAFETY TIP

Model Magma should always be observed and investigated inside the plastic box to avoid dangerous situations that could result from spillage.

Inquiry 11.1 continued

5 Working inside the plastic box at all times:

A. *Very gently* place a container of soil into the container of Model Magma™, as shown in Figure 11.1.

B. Then hold the container in both hands. *Slowly* press down on the container of soil.

C. Press down for only a few seconds. Then stop!

D. What do you observe? Record your observations. You can use both words and pictures.

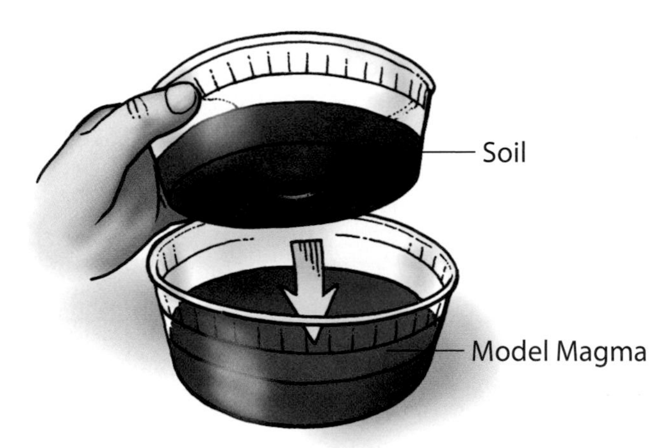

CAREFULLY PLACE THE CONTAINER OF SOIL OVER THE MODEL MAGMA.
FIGURE **11.1**

6 Again, working *very slowly*:

A. Push the container of soil farther down into the container of Model Magma, until you cannot press down any more.

B. Discuss your observations with your group and record your findings.

C. Then set aside the stacked Model Magma and soil.

SAFETY TIPS

Be careful when handling the heated Model Magma. If you must collect it yourself, use the beaker clamp, as shown in Figure 11.2.

Fill your beaker at the area where the hot pots are set up. Do not remove the hot beaker from that area.

USE A BEAKER CLAMP WHEN HANDLING A HOT BEAKER.
FIGURE **11.2**

7. Have one member of your group collect the hot Model Magma from your teacher. Fill your empty plastic container (the one with no hole) about three-fourths full of the hot Model Magma.

8. Stir the heated Model Magma with the spoon. Do *not* touch the magma with your hands. It is hot. How is the heated Model Magma different from the room-temperature Model Magma? Record your general observations of it. How do you think the heated Model Magma will behave under the soil? Discuss this with your group.

9. Repeat Procedure Steps 5 and 6 using the heated Model Magma and a new container of soil. Discuss your observations with your group after each step. Record your findings.

10. Slowly lift the container of soil out of the container of heated Model Magma, as shown in Figure 11.3. Hold the soil container above the container of magma. Observe what happens to the magma and soil. Record your observations.

LIFT THE CONTAINER OF SOIL OUT OF THE CONTAINER OF HEATED MODEL MAGMA.
FIGURE **11.3**

Inquiry 11.1 continued

11. Clean up by following your teacher's instructions.

12. Answer these questions:

    A. Were there any signs in the soil that magma was moving under the ground?

    B. How does rising magma affect land with no hard rock above it?

    C. What happened when the room-temperature Model Magma™ reached the surface of the soil?

    D. How did the flow of the room-temperature Model Magma differ from the flow of the heated Model Magma?

    E. What happened to the soil when you drained the heated magma from it?

INQUIRY 11.2

INVESTIGATING LAVA AND NEW LANDFORMS

PROCEDURE

1. Why and how do you think rock melts? Share your ideas with the class.

2. Obtain your plastic box of materials. Remove the block of wax and the hand lenses. Examine the wax. What are its properties? For example, what are its color, size, shape, and odor?

3. Share your observations about the wax.

4. Make a prediction. What do you think would happen if you heated the wax?

SAFETY TIPS

Be very careful with the hot wax. Keep the lid on the container until you are ready to use the wax.

If your teacher does not distribute the wax, you will go to the hot pot area to collect it yourself. Your teacher will demonstrate how to use a clamp to hold the hot beaker. Do not carry a beaker of wax across the room.

During cleanup, do not place the wax into the water bath. Your teacher will put it back in the water bath so it can be reused by other classes.

5. Look at the heated wax that your teacher shows you. What are its properties now? How are they the same as the properties of the solid wax? How are they different? Like the solid wax, solid rock will also melt, but at much higher temperatures. In this inquiry, you will use melted wax to model lava flow on land and in water.

6. After reviewing the Procedure and the Safety Tips, design a table in your science notebook to record your results in an organized way.

7. Remove the remaining materials from the plastic box and create a slope using your box lid. Cover the slope with waxed paper, as shown in Figure 11.4. The waxed paper should extend past the lid and flatten out on the table, as shown. (Use your piece of tape if the waxed paper will not stay in place on the lid.)

8. Discuss with your group how you think melted wax will change when you pour it onto the *flat* part of the waxed paper. Predict how the melted wax will behave when you pour it onto the *sloping* part of the waxed paper.

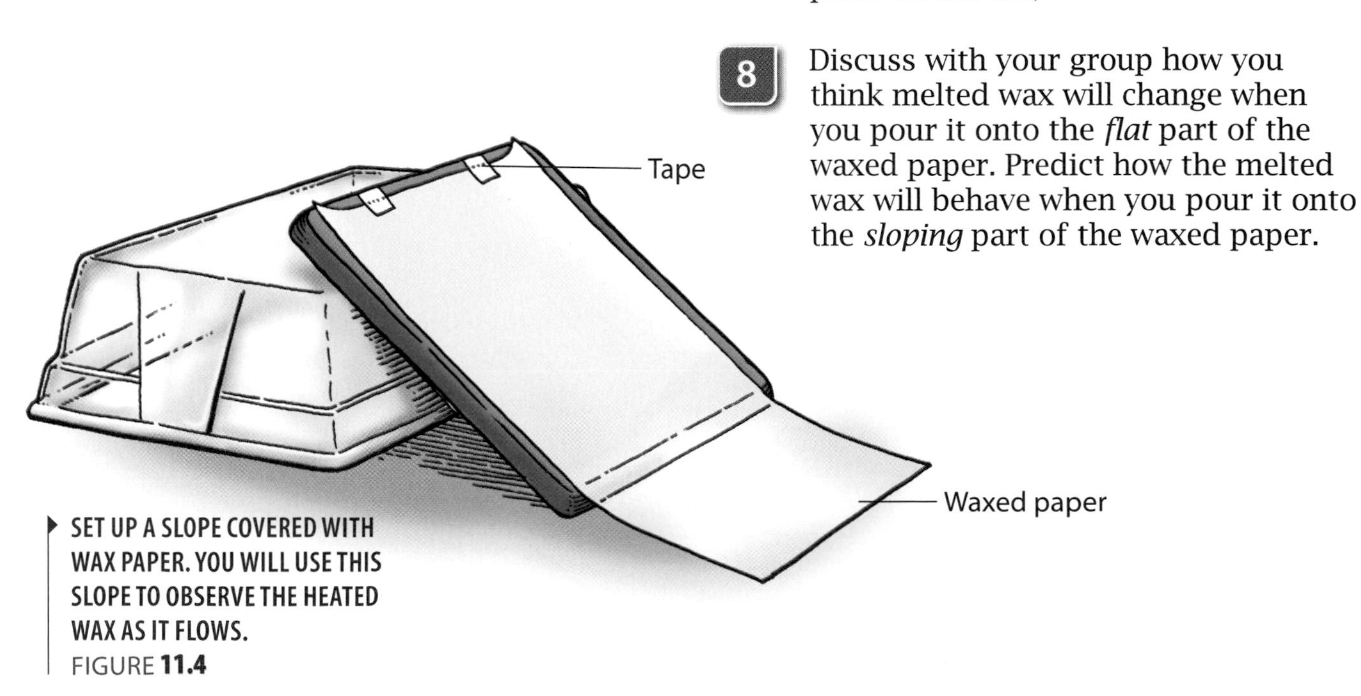

SET UP A SLOPE COVERED WITH WAX PAPER. YOU WILL USE THIS SLOPE TO OBSERVE THE HEATED WAX AS IT FLOWS.
FIGURE **11.4**

Inquiry 11.2 continued

9 Obtain one container of melted wax from your teacher. Test your predictions by doing the following:

A. Pour one-fourth of the container of hot wax onto the *flat* part of the waxed paper. Notice how the melted wax behaves.

B. Pour one-fourth of the container of hot wax onto the *sloping* part of the waxed paper, as shown in **Figure 11.5**. Discuss your observations with your group.

C. Wait 5 to 10 seconds. Then pour another one-fourth of the hot wax onto the *same* area of the sloping waxed paper. Wait again. Then pour the final one-fourth of the hot wax onto the same sloping area of the waxed paper. Discuss your observations with your group, then record your findings.

D. Use a hand lens to look carefully at the wax. What is its texture? What is its appearance? Is it a solid or a liquid? Record your observations.

CAREFULLY POUR THE WAX ONTO THE SLOPING PART OF THE WAXED PAPER.
FIGURE **11.5**

10 Now get a second container of heated wax while another member of your group gets 200 mL of very cold water from your teacher.

11 What do you think will happen when the melted wax flows into cold water? Discuss your predictions with your group.

12 Test your predictions. Kneel down so that you are at eye level with the beaker. Place an index card behind the beaker so you can see clearly. Quickly pour the entire container of hot wax into the beaker of cold water. Do not touch the wax for several minutes. Record your observations. After several minutes, remove the wax from the water. Turn it upright. Draw a picture of what happened to the "undersea lava."

13 Clean up by following these steps:

A. Return any solid wax to your teacher.

B. Pour out the water in your beaker.

14 Answer the following questions based upon what you have observed in this inquiry:

A. Describe the movement of the melted wax on the slope. How did that movement differ from the movement on the flat part of the waxed paper?

B. Describe the texture of the cooled wax.

C. How did the behavior of the cooling wax on the waxed paper compare with its behavior in the water?

15 Apply what you observed in this inquiry to cooling lava on the earth by answering these questions:

A. How do you think lava forms rock?

B. How do you think volcanic mountains (such as Mt. St. Helens) and volcanic islands (such as Hawaii and Iceland) form?

C. Under what circumstances do you think lava flows into the ocean?

D. What happens to lava when it flows into the ocean or erupts onto the ocean floor?

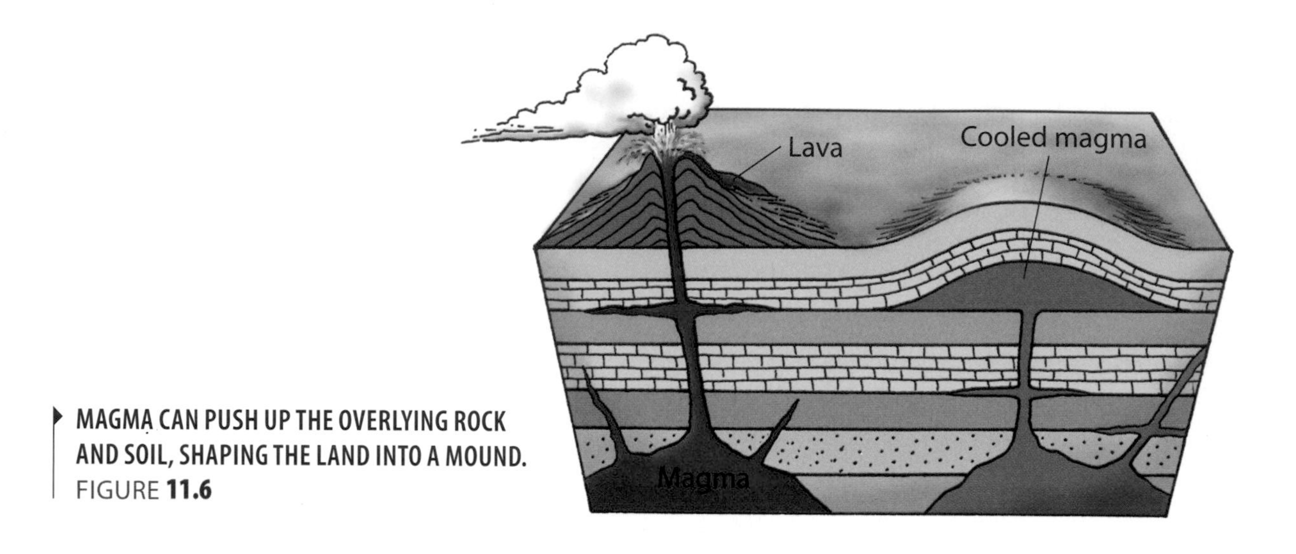

MAGMA CAN PUSH UP THE OVERLYING ROCK AND SOIL, SHAPING THE LAND INTO A MOUND.
FIGURE **11.6**

THIS PHOTO SHOWS A LAVA DOME LOCATED AT THE TOP OF THE NOVARUPTA VENT IN THE VALLEY OF TEN THOUSAND SMOKES IN KATMAI NATIONAL PARK AND PRESERVE, ALASKA. THE BULBOUS, STEEP-SIDED DOME FORMED AT THE TOP OF THE VOLCANO WHEN THICK, RELATIVELY COLD MAGMA (THAT DID NOT FLOW EASILY) CAME OUT OF THE VOLCANIC OPENING. IN MOST CASES, A LAVA DOME WILL GROW AS SUCCESSIVE ERUPTIONS ADD TO ITS SHAPE.
FIGURE **11.7**

PHOTO: U.S. Geological Survey/photo by T. Miller

IF LAVA IS RUNNY, IT MAY FLOW QUICKLY OVER THE SURFACE OF THE EARTH AND COVER A WIDE AREA, AS SHOWN HERE. THIS IS CALLED A LAVA FLOW. LIKE THE FLOW OF A RIVER, LAVA FLOW FOLLOWS THE PATH OF LEAST RESISTANCE—FOR EXAMPLE, INTO GULLIES AND VALLEYS. THEREFORE, SCIENTISTS CAN PREDICT WHERE THE LAVA WILL TRAVEL. PREDICTING THE PATH OF A LAVA FLOW CAN HELP SAVE LIVES. THIS PICTURE OF A LAVA FLOW WAS TAKEN FROM THE PU'U 'O'O-KUPAIANAHA ERUPTION IN HAWAII.
FIGURE **11.8**

PHOTO: U.S. Geological Survey

OVER TIME, THE SURFACE OF A LAVA FLOW COOLS AND HARDENS INTO NEW ROCK. THE COOLED LAVA IN THIS PHOTO CRACKED WHEN HOT LAVA FLOWING BENEATH IT PUSHED UP ON ITS BRITTLE SURFACE.
FIGURE **11.9**

PHOTO: Pierre Guinoiseau/creativecommons.org

SOMETIMES MAGMA RETREATS OR ERUPTS FROM A SHALLOW, UNDERGROUND MAGMA CHAMBER. WITHOUT THE MAGMA TO SUPPORT THE GROUND ABOVE IT, THE OVERLYING ROCK COLLAPSES AND FORMS A LARGE, STEEP-SIDED, CIRCULAR OR OVAL VOLCANIC DEPRESSION, CALLED A CALDERA. (A CALDERA IS NOT A CRATER, WHICH IS SMALLER AND FORMS WHEN ROCK EXPLODES FROM THE VOLCANO DURING AN ERUPTION.)
FIGURE **11.10**

PHOTO: U.S. Geological Survey

SOME LAVA COOLS TO FORM A SKIN OR BLACK BASALT ROCK THAT IS WRINKLED BY THE HOT LAVA STILL FLOWING UNDER IT. IT HAS A SMOOTH, ROPY TEXTURE AND IS TYPICAL OF THE LAVA FLOW IN HAWAII.

PHOTO: U.S. Geological Survey/photo by J.D. Griggs

FIGURE **11.11**

SOMETIMES LAVA OOZES INTO THE OCEAN AND COOLS ON CONTACT WITH THE WATER. THIS IS ONE WAY THAT PILLOW LAVA FORMS.

PHOTO: U.S. Geological Survey/photo by J.D. Griggs

FIGURE **11.12**

REFLECTING ON WHAT YOU'VE DONE

1. Use your own observations from the inquiries to develop working definitions for the terms "magma" and "lava." Record your definitions in your science notebook.

2. Think about what happened to the soil when the Model Magma™ moved under it. Then answer these questions:

   A. Why do you think an earthquake almost always happens before lava erupts onto the land?

   B. What are some of the signs that help volcanologists predict a volcanic eruption?

3. Apply what you observed in this lesson to magma and lava on the earth. Look at Figures 11.6 through 11.10. Read the captions. For each figure, describe what you did during Inquiry 11.1 that helped you better understand how each landform was made. Record your ideas in your science notebook.

4. Look at the photographs in Figures 11.11 and 11.12. Read the captions. For each one, describe what you did during Inquiry 11.2 that helped you better understand how each landform was created. Record your ideas in your science notebook.

5. Look ahead to the next lesson, in which you will examine how lava can form different sizes and shapes of volcanoes.

READING SELECTION

EXTENDING YOUR KNOWLEDGE

AN ISLAND IS BORN

The explosion of an underwater volcano 32 kilometers (20 miles) south of Iceland in November 1963 gave scientists a rare chance to observe the formation of new land—the island of Surtsey. The first sign of island formation was smoke emerging from the North Atlantic. A fisherman thought this smoke was a ship in trouble, but it was actually the birth of an island.

Surtsey was formed by a volcano that erupted in relatively shallow water, about 130 meters (427 feet) deep. Like Iceland, Surtsey formed over an unusually hot region of the mantle, called a hot spot. A hot spot is a localized zone of melting in the mantle that is fixed in the mantle under the plate. Volcanoes form above the hot spot. The Hawaiian Islands also formed over a hot spot.

The hot spot below Surtsey is located in a region that happens to coincide with the Mid-Atlantic Ridge. This is why a great volume of lava flows from the vents of Iceland. During the Surtsey eruption, hot magma shot upward into the ocean waters and outward in horizontal blasts, causing the island to grow out as well as up. The final eruptions were lava flows that were harder and more compact than earlier ash deposits. In just a few weeks, the lava flows had created a hard crust that protected the island from washing away immediately, and an island was born. ■

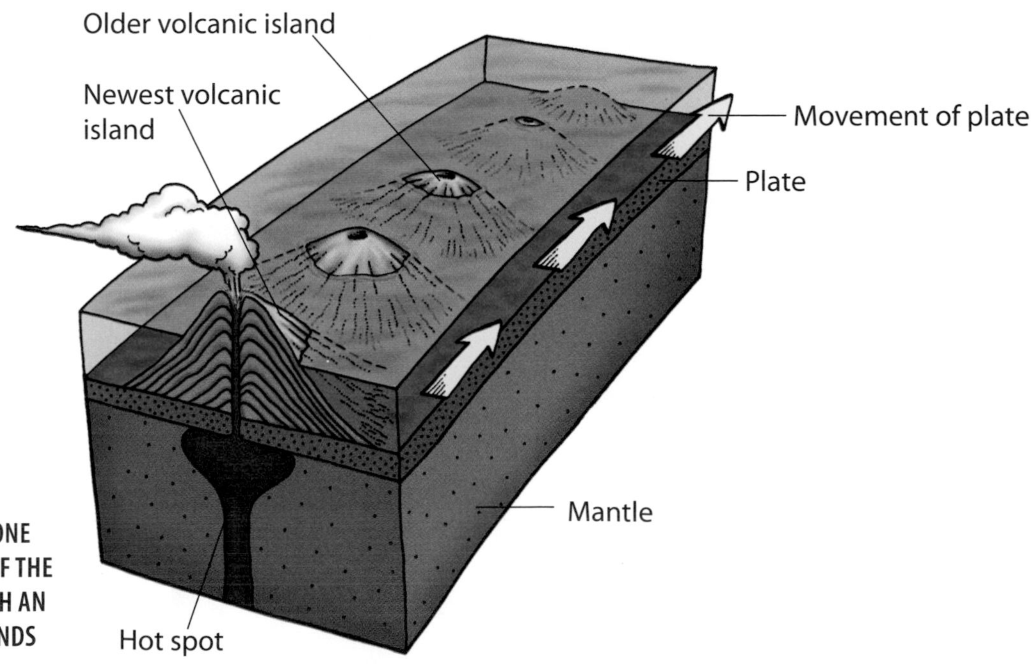

A HOT SPOT IS A LOCALIZED ZONE OF MELTING IN THE MANTLE. IF THE HOT SPOT IS LOCATED BENEATH AN OCEAN PLATE, VOLCANIC ISLANDS CAN FORM.

READING SELECTION

EXTENDING YOUR KNOWLEDGE

THE LAVA THAT FLOWED DURING THE SURTSEY ERUPTION BEGAN FROM A SHALLOW, UNDERWATER VOLCANO.

PHOTO: University of Colorado, Boulder, CO/National Geophysical Data Center/NOAA

WITHIN A FEW DAYS, THE NEW ISLAND OF SURTSEY HAD FORMED.

PHOTO: Photographer Howell Williams/National Geophysical Data Center/NOAA

DISCUSSION QUESTIONS

1. Like Surtsey, the Hawaiian Islands are formed by a hot spot. As they drift away from the hot spot because of the movements of the Pacific plate, what do you think happens to them?
2. Other than from hot spot volcanoes, what are some other ways that islands can form? Use Internet and library resources to gather additional information, if necessary.

READING SELECTION

EXTENDING YOUR KNOWLEDGE

Volcanologists Talk About Their Work

A VOLCANOLOGIST COLLECTS SAMPLES FROM A STILL-MOLTEN LAVA FLOW.

PHOTO: U.S. Geological Survey

Q: What is it like to work on volcanoes?

A: Volcanoes are beautiful places where forces of nature combine to produce awesome events and spectacular landscapes. For most of us, they are also fun to work on! There's something moving about the idea of magma rising from deep inside our restless planet to flow gracefully onto its surface, as in Hawaii, or to explode violently into its atmosphere, as at Mt. St. Helens. As one scientist put it, "I'm fascinated by the knowledge that some of the gases I breathe were once miles deep in the earth and arrived in my lungs by way of a volcano." Perhaps no spot on earth is untouched by the effects of volcanoes. In fact, more than half of the earth's surface is covered by volcanic flows, especially the seafloor. All forms of life on earth are linked in some way to volcanic activity. With this in mind, what could be more exciting or rewarding than to work on an active volcano?

Q: Are you scared when you work on an active volcano?

A: "Excited" is the first word that comes to mind when most of us think about our work at active volcanoes. Safety is always our primary concern, because volcanoes can be dangerous places. But we manage personal risk in the same way as police officers, astronauts, or those in any other hazardous professions. We try hard to understand the risk that is built into any situation. Then we train and equip ourselves with the right tools and support to be safe. Such training involves learning the past and current activity of the volcano, first aid, helicopter safety procedures, and wilderness survival techniques.

READING SELECTION

EXTENDING YOUR KNOWLEDGE

VOLCANOLOGISTS REPAIRING EQUIPMENT AT A SEISMIC STATION ON REDOUBT VOLCANO, ALASKA

PHOTO: U.S. Geological Survey/Alaska Volcano Observatory/photo by Game McGimsey

Some of us, however, have experienced situations that were more than exciting. In the words of one scientist, "Scared? Oh sure. When a little steam explosion occurred from the dome at Mt. St. Helens in 1982, three of us were surveying the dome from less than 100 meters (328 feet) away. As soon as we saw the basketball size rocks streaming through the air, we ran for cover beneath a huge block of ice on the crater floor. Until the rocks stopped landing all around us, I was absolutely terrified."

Q: How about when the volcano is showing signs of activity and you have concluded the volcano is likely to erupt soon?

A: This is the most anxious time, because generally there is nothing more to be done than wait, watch, and hope that your team is right in its assessment of the situation. With modern monitoring instruments, an active volcano can seem almost overwhelming at this stage. Earthquakes can happen virtually nonstop for hours or days. Swelling or cracking of the ground occurs at rates that keep going up and up. And changes happen in the kinds and amounts of volcanic gases being released. Even so, there are always uncertainties, including the very real possibility that the process will simply stop before magma reaches the surface, and you will be asked to explain why there was so much fuss over a "failed eruption."

Q: What precautions do scientists take?

A: Restless volcanoes can be very dangerous places, but it is possible to work safely around them if you are properly prepared. First and foremost, scientists protect themselves by working as a team to create a safety net in which all the important bases are covered. Like a professional driving team, a volcano-response team includes key staff who know the monitoring equipment extremely well. They include experts in several scientific

IN OREGON, A SCIENTIST GENERATES ELECTRONIC DISTANCE MEASUREMENT (EDM) LINES WHILE THE HELICOPTER MEASURES AIR TEMPERATURE ALONG THOSE LINES TO MONITOR VOLCANOES FOR POTENTIAL ERUPTIVE ACTIVITY. THESE TECHNIQUES CAN DETECT MINOR SWELLING AND HEATING THAT OFTEN PRECEDE ERUPTIONS.

PHOTO: U.S. Geological Survey/Cascades Volcano Observatory/photo by Lyn Topinka

disciplines who can interpret data from the field.

Q: What education do you need to become a volcanologist?

A: There are many paths to becoming a volcanologist. Most volcanologists have a college or graduate school education in a scientific or technical field, but the range of specialties is very large. Training in geology, geophysics, geochemistry, biology, biochemistry, mathematics, statistics, engineering, atmospheric science, remote sensing, and related fields can be applied to the study of volcanoes and the interactions between volcanoes and the environment. The key ingredients are a strong fascination and boundless curiosity about volcanoes and how they work. From there, the possibilities are almost endless. ■

DISCUSSION QUESTIONS

1. What qualities must a volcanologist have in order to be successful?
2. Do you think it is possible to work on the science of volcanoes without walking on active volcanoes? If so, how?

LESSON 12

VISCOSITY AND VOLCANO TYPES

THIS PHOTO, TAKEN IN 1947 IN PARICUTÍN, MEXICO, SHOWS AN ERUPTION OF THE PARICUTÍN VOLCANO AT NIGHT. GLOWING HOT, BROKEN ROCKS OUTLINE THE SHAPE OF THE VOLCANO, CALLED A CINDER CONE.

PHOTO: U.S. Geological Survey

INTRODUCTION

When a volcano erupts, a mixture of red-hot lava, rocks, and gases burst into the atmosphere. Some volcanoes spew runny lava onto the earth's surface; this lava flows freely over wide areas. Other volcanoes ooze sticky lava that flows only a short distance.

In Lesson 11, you learned that magma is melted rock beneath the earth's surface, and that when it reaches the earth's surface, it is called lava. The way in which magma and lava flow and whether fragments of lava and rock erupt from the volcano affect a volcano's shape and size. In this lesson, you will design an investigation to test how liquids flow and the conditions under which this flow changes. You will then relate your observations to lava flow and volcano type.

OBJECTIVES FOR THIS LESSON

- Classify images of volcanoes on the basis of observable properties, including shape and size.
- Identify and compare how three different liquids flow.
- Observe changes in how two liquids flow when they are mixed with a solid.
- Observe changes in how two liquids flow when they are heated.
- Develop a working definition for the term "viscosity."
- Relate the viscosity of lava to the type of volcano formed, and model the formation of each type.

MATERIALS FOR LESSON 12

For your group

- 1 copy of Student Sheet 12.1a: Investigating the Viscosity of Liquids—Planning Sheet
- 1 copy of Student Sheet 12.1b: Investigating the Viscosity of Liquids—Group Data and Observations
- 1 set of 9 Volcano Cards

Sample substances:

- 1 cup of dark corn syrup, with lid
- 1 cup of shampoo, with lid
- 1 cup of water, with lid
- 1 cup of sand, with lid

- 1 plastic box with lid
  - 1 stopwatch
  - 1 ruler
  - 2 sheets of waxed paper
  - 1 measuring spoon
  - 2 coffee stirrers
  - 2 empty plastic cups

Paper (to cover work areas)

GETTING STARTED

1. Review Lesson 11 with your class by describing the difference between magma and lava. With the class, discuss how you think the properties of magma and lava might affect a volcano's shape or size.

2. Collect one set of Volcano Cards from your teacher. Work with your group to classify the volcanoes on the basis of their observable properties, such as shape and size.

3. Share your classifications with the class. Explain why you classified the volcanoes as you did. Return your cards to the teacher.

4. Why do you think volcanoes have different shapes and sizes? Discuss this with your class. To test how the flow of lava and magma affect volcano type, you will investigate the flow of three different liquids.

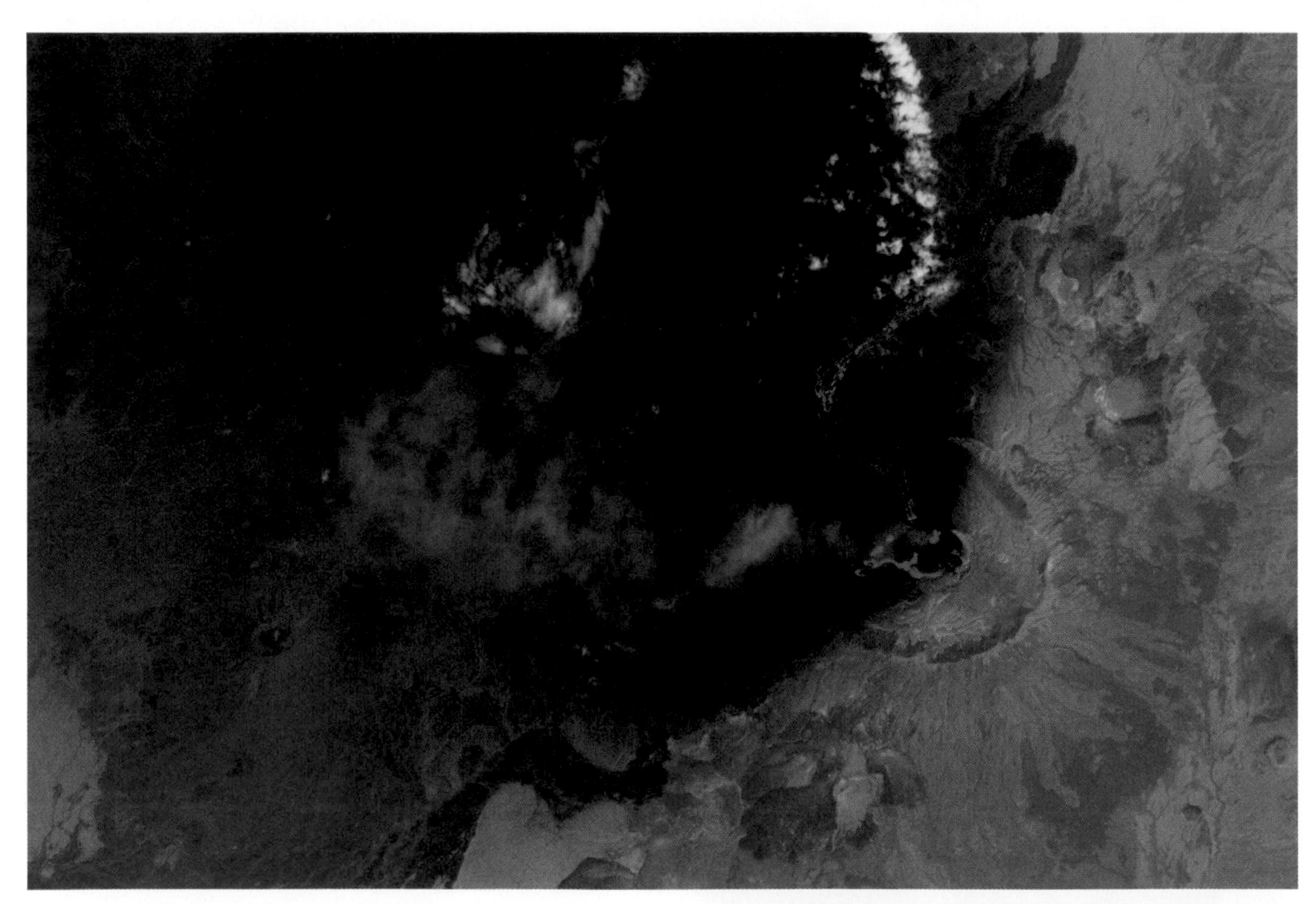

LAVA FLOWS FROM THE NABRO VOLCANO IN EAST AFRICA. THIS VISIBLE AND INFRARED LIGHT IMAGE SHOWS HOT LAVA AS ORANGE-RED AND COOLING LAVA FADING TO BLACK.

PHOTO: NASA Earth Observatory image by Robert Simmon, using EO-1 ALI data

INVESTIGATING VISCOSITY AND VOLCANO TYPE

PROCEDURE

1. Look at the three liquids selected for Inquiry 12.1: corn syrup, shampoo, and water. What general observations can you make about each liquid? Discuss them with the class. Then make some predictions by answering these questions:

   A. How do you think each liquid will flow at room temperature?

   B. How do you think the room-temperature liquids would flow if you added sand to them?

   C. How do you think the liquids would flow if they were heated?

2. Look at the materials you will use for the investigation. Groups will test different combinations of substances. At the end of the inquiry, you will share your data with the class.

3. Before you design your investigation, answer the following questions with your class:

   A. The purpose of the experiment is to compare how fast or slow each of three liquids flows. Given these materials, how would you set up an investigation to test the flow of each liquid at room temperature?

   B. What things would you need to keep the same in each setup?

   C. Each group will conduct a different part of the investigation, and then the data will be combined. Why is it important that all groups agree to use the same setup and procedures?

   D. How could you determine the effects of heat on the flow of a liquid?

   E. How could you determine whether adding a solid, such as sand, to the liquid affects the rate at which it flows?

Inquiry 12.1 continued

4. Your group will test one of the combinations below:

   A. Room-temperature corn syrup versus room-temperature shampoo

   B. Room-temperature water versus room-temperature corn syrup versus room-temperature shampoo

   C. Room-temperature corn syrup versus heated corn syrup

   D. Room-temperature shampoo versus heated shampoo

   E. Room-temperature shampoo versus room-temperature shampoo mixed with sand

   F. Room-temperature corn syrup versus room-temperature corn syrup mixed with sand

5. Complete Student Sheet 12.1a: Investigating the Viscosity of Liquids—Planning Sheet. Be sure to answer the following questions as you proceed with the setup:

   A. What question will your group try to answer?

   B. What do you think will happen when you heat the liquid or add a solid to your liquid?

   C. What materials and procedures will you use?

   D. What things must you keep the same when testing each liquid?

   E. What will you measure, and how?

   F. How many trials should you conduct to determine an average?

   G. Discuss this with your class.

6. Complete Student Sheet 12.1b: Investigating the Viscosity of Liquids—Group Data and Observations.

7. Make certain your group's box of materials contains the substances that your group has been assigned to test.

8. Conduct the experiment and record your findings. Keep in mind the following points:

   A. Each plastic box contains enough materials for several classes. You will not use all of the materials in this period.

   B. Try to work inside your plastic box. You may also want to put a piece of paper on your work space for easy cleanup.

   C. If your group is testing a heated liquid, use your empty plastic cups to collect, under your teacher's supervision, the needed heated substance from the hot pot area.

   D. If your group is testing a mixture of sand and room-temperature liquid, use the plastic cups and stirrers to mix the sand and liquid. Make certain the volume of the sand and liquid mixture equals the volume of the liquid only.

9. When your group is finished, follow your teacher's directions for cleanup.

REFLECTING ON WHAT YOU'VE DONE

1. Share your observations and data with the class. Your teacher will combine data from all groups into one data table.

2. Answer the following questions in your science notebook:

   A. Which room-temperature liquid flowed the fastest? Which flowed the slowest? What evidence do you have to support your answer?

   B. How did adding sand to each liquid change how it flowed? Support your answer with evidence.

   C. How did heating each liquid change how it flowed? Support your answer with evidence.

3. Write a working definition of "viscosity." Then explain how you think the viscosity of lava affects the shape of a volcano.

4. Look again at the Volcano Cards. Then answer these questions:

   A. What is the shape of a volcano that forms from fast-flowing lava? Which Volcano Cards show this type of volcano?

   B. What is the shape of a volcano that forms from viscous, slow-moving lava? Which Volcano Cards show this type of volcano?

5. A lava's viscosity is one factor affecting the shape and size of the volcano. Read "Volcano Types" on pages 166–169 to find out more.

6. With your group, categorize each of the Volcano Cards as a shield volcano, composite volcano, or cinder cone. Use the reading selection "Volcano Types" to help you.

7. Watch as your teacher models the formation of the three types of volcanoes. Compare the model volcanoes with the photographs in the reading selection and on the cards.

8. Discuss the advantages and disadvantages of these models.

READING SELECTION

EXTENDING YOUR KNOWLEDGE

VOLCANO TYPES

Scientists classify volcanoes into three basic categories on the basis of shape and size. Let's take a closer look at these types of volcanoes.

SHIELD VOLCANOES

While many people think of volcanic eruptions as being explosive, many volcanic areas produce quiet, oozing lava. Fissures and hot spots are two examples. Fissures are long fractures in the earth's crust. Instead of erupting from one central vent, or opening, lava erupts gently like a fountain from the fissure in a long line. Fissures normally form in areas where two plates separate, such as along a mid-ocean ridge.

Like fissures, hot spots produce quiet eruptions. Most hot spots form under a plate instead of along its boundaries. Other hot spots coincide with mid-ocean ridges. Both fissures and hot spots produce a runny lava that spreads out to form a wide, broadly sloping volcano. These volcanoes are called shield volcanoes because they resemble a warrior's shield. The slopes of a shield volcano are rarely steep; these volcanoes are wide and flat. Over thousands of years, shield volcanoes can reach massive size, for example, 9 kilometers (5.6 miles) high and 193 kilometers (120 miles) wide. The Hawaiian Islands and Iceland are examples of shield volcanoes.

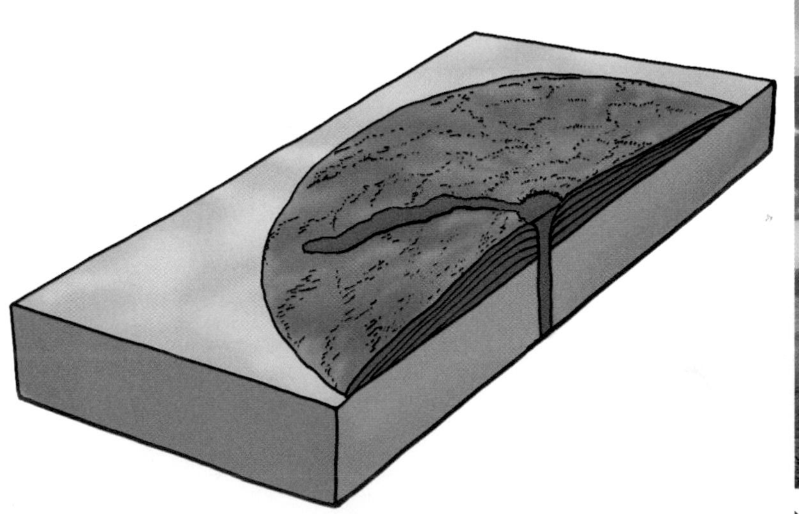

THIS CROSS-SECTION SHOWS HOW REPEATED LAYERS OF FLOWING LAVA FORM SHIELD VOLCANOES.

SHIELD VOLCANOES ARE FLAT ON TOP AND BROADLY SLOPING. THIS PHOTO SHOWS MAUNA LOA, A SHIELD VOLCANO IN HAWAII.

PHOTO: U.S. Geological Survey/photo by D. Little

COMPOSITE VOLCANOES

Composite volcanoes are tall and pointed. Th are some of the most picturesque volcanoes in the world because of their height and snow capped summits. They form from alternating eruptions of lava and ash. A composite volcai is flat toward the base and steep toward the summit. The lava is sticky and does not flow before it solidifies.

These tall volcanoes usually form where two plates collide and one overrides the othe Thick magma and water from the sinking oceanic plate cause the volcano to be explosi Water dissolves within the magma and travel upwards as small bubbles, like the bubbles in a carbonated soft drink. When the magma explodes from the volcano, it breaks the lava and rocks along the vent into pieces. Alternating layers of lava and fragmented rock pile up. Mt. St. Helens in Washington state and Mt. Fuji in Japan are composite volcanoes.

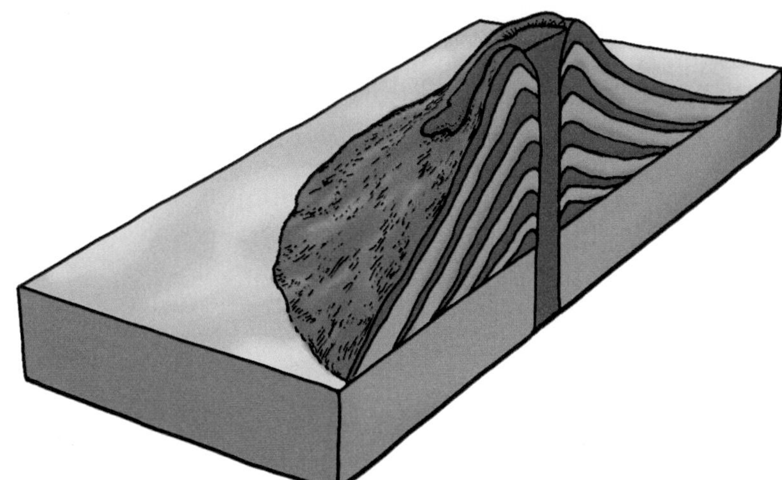

THIS CROSS-SECTION SHOWS HOW ALTERNATING LAYERS OF CINDERS AND LAVA FORM A COMPOSITE VOLCANO.

COMPOSITE VOLCANOES FORM FROM ALTERNATING ERUPTIONS OF LAVA AND ASH. THIS PHOTO SHOWS MT. MAGEIK VOLCANO AS SEEN FROM THE VALLEY OF TEN THOUSAND SMOKES, KATMAI NATIONAL PARK AND PRESERVE, ALASKA.

PHOTO: U.S. Geological Survey/photo by R. McGimsey

READING SELECTION

EXTENDING YOUR KNOWLEDGE

CINDER CONE VOLCANOES

Cinder cone volcanoes are smaller than shield and composite volcanoes. If the eruption contains thick magma, the gas pressure shatters the rock within the volcano into small pieces. In other cases, the lava in the air may harden and fall as fragments. These small pieces are called cinders. These cinders accumulate around the opening, or vent, of the volcano. These volcanoes tend to be explosive, which is why the rock breaks into fragments. Cinder cones can also ooze lava at the base of the cone. Eldfell in Iceland and Sunset Crater are cinder cone volcanoes. ■

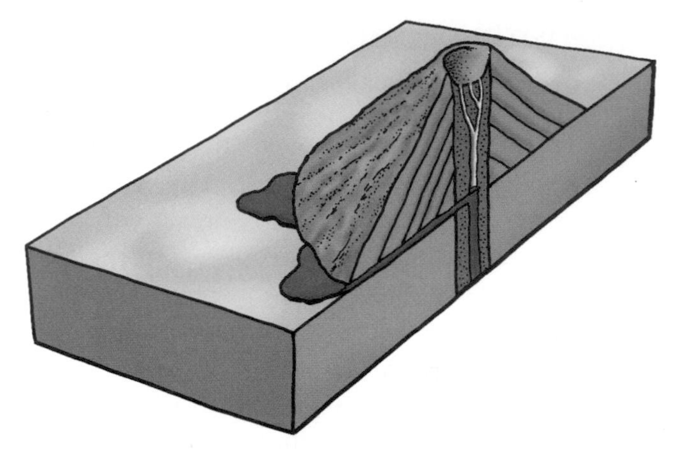

THIS CROSS-SECTION SHOWS HOW LAYERS OF CINDER FORM A CINDER CONE.

CAPULIN MOUNTAIN IN NEW MEXICO IS A HUGE CINDER CONE VOLCANO. ITS CONE IS MORE THAN 300 METERS (ABOUT 1000 FEET) ABOVE ITS BASE.

PHOTO: U.S. Geological Survey/photo by R.D. Miller

LAVA IN MOTION

What is 61 meters (200 feet) across, moves up to 64 kilometers (40 miles) per hour, and can destroy 14 villages?

LAVA!

On January 17, 2002, people living in the city of Goma in the Republic of Congo awoke to a startling reality. The sky was dark with falling ash, and three streams of red-hot lava were racing down the slopes of Mount Nyiragongo towards their homes. All the people could do was get out of their houses and run until they crossed the nearby Rwandan border. While the volcanic event lasted only two days, thousands of people returned to find their homes burned to the ground.

The Nyiragongo volcano is a stratovolcano (also called a composite volcano), the same type as Mt. St. Helens and Eyjafjallajökull. It's a steep-sided type of volcano that normally produces a viscous, slow-flowing lava, rich in silica. The Nyiragongo volcano, however, happens to have unusually thin, fast-flowing lava. At the top of Nyiragongo is a crater containing a lake of steaming lava; in the 2002 disaster, leaks in the side of the volcano widened into three fissures through which the pooled lava poured out.

The Nyiragongo volcano also erupted in 1977 and 1994. Given the timing of these events, what is your prediction for the date of the next eruption?

A PHOTO TAKEN BY SATELLITE SHOWS THE PLUME OF ASH SHOOTING SKYWARD AS THE NYIRAGONGO VOLCANO ERUPTED.

PHOTO: NASA Goddard Space Flight Center

DISCUSSION QUESTIONS

1. Why are composite volcanoes so tall?
2. What could make some lava more viscous than other lava?

READING SELECTION

EXTENDING YOUR KNOWLEDGE

The Volcano Lovers

French geologists Maurice and Katia Krafft loved watching volcanoes erupt. They were so fascinated by these powerful forces of nature that watching volcanoes became their life's work. For over 20 years, they witnessed more than 140 eruptions—on every continent except Antarctica.

The Kraffts not only watched these eruptions, but they also took close-up pictures of them. They put themselves at great risk by getting close to volcanoes to understand them better. They knew the dangers of the fiery molten lava and scalding clouds of ash.

Maurice and Katia met during the 1960s while they were studying geology at the same university in France. Their passion for volcanoes brought them together. In 1968, after they were married, they founded a center for volcanology. ("Volcanology" is the study of volcanoes.) Their goal was to take measurements of molten lava, to analyze volcanic gases, and to record volcanic eruptions on film.

Through books and lectures, the Kraffts raised enough money to support their expeditions. As soon as they heard about an eruption anywhere in the world, they packed and boarded the next plane. In an average year, they visited about three big eruptions. In 1988, when volcanoes seemed to be erupting everywhere, the Kraffts circled the globe several times.

The films made by the Kraffts reveal the beauty and power of volcanic eruptions. The Kraffts picked up useful details that helped geologists understand volcanoes better. For example, they filmed molten lava that was black, instead of red, flowing from a volcano in the African country of Tanzania. No one knew about this kind of lava, which came from rock that melts at 500°C (932°F). The more common red lava comes from rock that melts at about 1000°C (1832°F).

Having witnessed hundreds of volcanic eruptions, the Kraffts were concerned about the danger of volcanoes when people living near them are not properly warned. For instance, when the Nevada del Ruiz volcano in Colombia, South America, started erupting, volcanologists advised authorities of the danger to people living in nearby towns. Authorities did not believe that people living 47 kilometers (29 miles) away were in danger, and the scientists' warnings were not heeded. As a result,

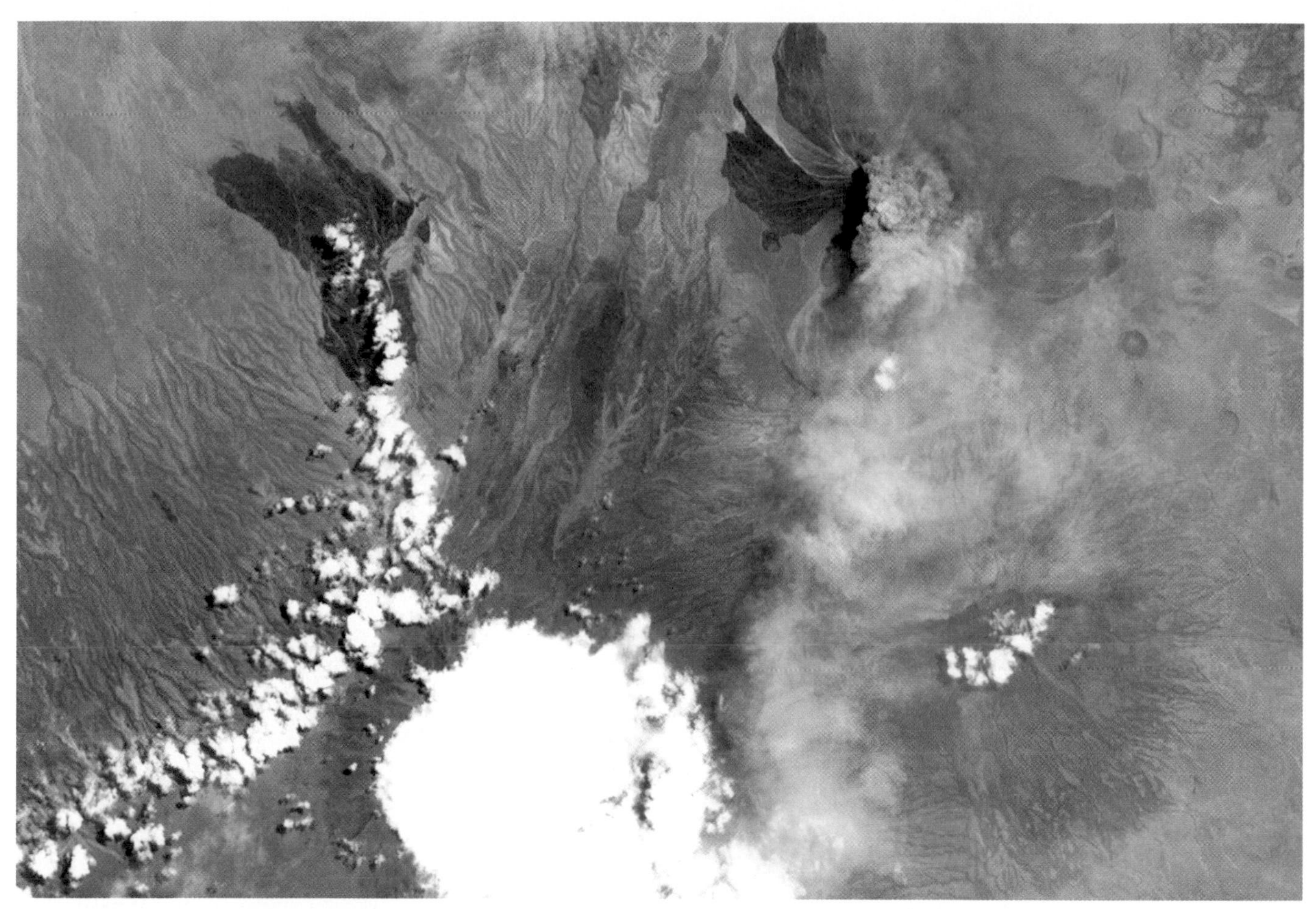

OL DOINYO LENGAI IN TANZANIA IS THE ONLY KNOWN ACTIVE VOLCANO IN THE WORLD THAT SPEWS BLACK LAVA.

PHOTO: NASA image created by Jesse Allen, using data provided courtesy of NASA/GSFC/METI/ERSDAC/JAROS, and U.S./Japan ASTER Science Team. Image interpretation by Greg Vaughan, Jet Propulsion Laboratory.

22,000 people died in the mudflows caused by a later eruption. Had these people walked 200 meters (656 feet) to the nearest hill, Maurice said, they would have lived.

The Kraffts and others felt something had to be done to protect the 500 million people who live near the world's active volcanoes, such as in Iceland. Maurice offered the best of his film footage, plus his knowledge of both volcanoes and cinematography, to make a video that would document the hazards of volcanoes.

In June 1991, Maurice and Katia went to Unzen volcano in Japan. A deadly mix of hot gas, ash, and rock was surging down the volcano's upper slopes 35 times a day. It was the perfect chance for the Kraffts to film the flows and educate officials about the danger. Each time a chunk of lava near the summit broke away and began tumbling down, it shattered. The rock slide became a rock-and-ash stream that careened down the mountain at up to 100 kilometers (62 miles) per hour.

READING SELECTION

EXTENDING YOUR KNOWLEDGE

HOUSES ALONG THE MIZUNASHI RIVER WERE DESTROYED BY MUDFLOWS CAUSED BY UNZEN VOLCANO IN NORTHER KYUSHU.

PHOTO: U.S. Geological Survey/photo by T.J. Casadeval

THIS AERIAL VIEW OF UNZEN SHOWS THE PATH OF PYROCLASTIC FLOW EXPELLED BY THE VOLCANO. IT WIPED OUT OR FLOODED A LARGE SECTION OF SHIMABARA, A TOWN ON THE COAST BELOW.

PHOTO: U.S. Geological Survey/photo by T. Kobayashi, Univ. Kagoshima

THE FIRST MAJOR ERUPTION OF MT. PINATUBO ON JUNE 12, 1991

PHOTO: U.S. Geological Survey/Cascades Volcano Observatory/photo by Richard P. Hoblitt

At the point when the mountain was sending down medium-sized flows, Maurice and Katia were witnessing quite a show. Suddenly, a huge chunk of the lava plunged toward them. They had no time to escape. The Kraffts, along with 49 other people, were killed.

Only two weeks later, because of the influence of the Kraffts' video, an evacuation saved an estimated 20,000 lives from a major eruption of Mt. Pinatubo, in the Philippine Islands. The work to which these two scientists devoted their lives continues to benefit people around the world. Through the films that the Kraffts made, the world can also share what they saw: the awesome beauty of a spectacular and deadly force of nature. ■

DISCUSSION QUESTIONS

1. Do you think it was wise for the Kraffts to have spent their lives documenting volcanoes, considering how dangerous it was?
2. Imagine you have been hired by the Japanese government to educate people about the dangers of area volcanoes. How will you do your job?

LESSON 13

IGNEOUS ROCK

THIS "HORNITO" IS MADE OF IGNEOUS ROCK. IT DEVELOPED ON KILAUEA VOLCANO IN HAWAII WHEN MOLTEN ROCK FORCED ITS WAY THROUGH AN OPENING. AS YOU WILL SOON DISCOVER, ITS APPEARANCE GIVES CLUES ABOUT HOW IT WAS FORMED AND WHAT IT IS MADE OF.

PHOTO: U.S. Geological Survey/photo by J.D. Griggs

INTRODUCTION

In some parts of the mantle, hot rock rises into the earth's crust and melts. When the melted rock eventually cools, either inside the earth or on the surface of the earth, it hardens. As it cools, its minerals form grains, which are called crystals. Rocks that form this way are called igneous, which comes from the Latin word for "fire."

In this lesson, you will examine five igneous rocks. You will analyze each rock's observable color, mineral composition, and texture (the size of its crystals). Your group will classify the rocks based on their properties.

OBJECTIVES FOR THIS LESSON

- Observe the properties of five igneous rocks.
- Sort and classify igneous rocks on the basis of color, mineral composition, and texture.

MATERIALS FOR LESSON 13

For your group

1 copy of Student Sheet 13.1: Observing the Properties of Igneous Rock
1 plastic box with lid
- 1 set of rocks, labeled #1 through #5
- 3 hand lenses
- 1 double-eye loupe (magnifier)
- 1 ruler

GETTING STARTED

1. Look at the sample rock your teacher shows you. Discuss the following with your class:

   A. What general observations can you make of this rock?

   B. How would you define the word "rock"?

   C. Do you think it would be possible for this rock to change into another rock? If so, how might this happen?

2. Look at Figure 13.1. Discuss its caption with your teacher. In this lesson, you will examine five rocks and the minerals that make them up.

THE GREEN ROCK YOU JUST OBSERVED IS MOSTLY MADE OF THE MINERAL OLIVINE, SHOWN HERE. OLIVINE HAS A CRYSTALLINE APPEARANCE, LIKE SALT. FOUR ELEMENTS—MAGNESIUM, IRON, SILICON, AND OXYGEN—CHEMICALLY COMBINE TO FORM OLIVINE. THE EARTH'S UPPER MANTLE CONTAINS A GREAT AMOUNT OF OLIVINE.
FIGURE **13.1**

PHOTO: NASA Ames Research Center, Tom Trower

OBSERVING IGNEOUS ROCK

PROCEDURE

1. Obtain a plastic box with five rock samples, hand lenses, a double-eye loupe, and a ruler. Using **Figure** 13.2 as a guide, talk with your teacher about how to use the double-eye loupe.

HOLD THE WIDE END OF THE LOUPE CLOSE TO YOUR EYE. THE SINGLE LOUPE WILL MAKE THE ROCK'S TEXTURE APPEAR 5 TIMES LARGER. THE DOUBLE LOUPE WILL MAKE IT APPEAR 12 TIMES LARGER.
FIGURE **13.2**

2. Spend a few minutes with your group examining and experimenting with the rocks and observation tools.

3. Share your initial observations of each rock with the class.

4. In this investigation you will do the following:

   A. You will observe each rock and record its observable color (that is, the variety of minerals you can see based on color) and its texture (that is, the size of its crystals or mineral grains). You will use your hand lens, loupe, and metric ruler; then you will classify the rocks into groups on the basis of these properties.

   B. How will you record your observations? Discuss your ideas with your class.

Inquiry 13.1 continued

5 Now complete and record your observations of the rock samples on Student Sheet 13.1: Observing the Properties of Igneous Rock. Be sure to do the following:

A. Describe the color of each sample.

B. Describe the texture of each sample by measuring the crystals under magnification.

C. Add any other observations.

D. Write a sentence or two to describe your classification system.

MICA
FIGURE **13.3**
PHOTO: U.S. Geological Survey/Mineral Collection of Brigham Young University Department of Geology, Provo, Utah/photo by Andrew Silver

QUARTZ
FIGURE **13.4**
PHOTO: BobMacInnes/creativecommons.org

PINK FELDSPAR
FIGURE **13.5**
PHOTO: Wikimedia Commons

REFLECTING ON WHAT YOU'VE DONE

1. Answer these questions. Use your observation table to support your answers.

   A. Do any of the rocks have similar properties? If so, describe those properties.

   B. Describe any rocks that do not share characteristics with the other rocks.

   C. What are some different ways to group the rocks? For example, consider the color or size of the mineral crystals. Explain why you grouped the rocks the way you did.

   D. Select one rock. Give it a name. Write the rock number next to the name. Be creative. Why did you give the rock that name?

2. Now look at the minerals in Figures 13.1, 13.3, 13.4, and 13.5. Use these photographs to identify some of the minerals that make up each of the rocks. List the minerals in your observation table.

3. Read "Igneous Rocks."

   A. What are the variables that determine the classification of igneous rocks?

   B. Which rocks do you think cooled slowly? Which rocks do you think cooled quickly?

4. Look ahead to Lesson 14, in which you will examine the formation of volcanic ash and the effect of ash fall from a volcanic eruption.

IGNEOUS ROCKS

Rocks formed by cooling magma or lava are called igneous rocks. Igneous rocks make up most of the ocean floor and continental crusts. These rocks are classified based on their texture, or crystal size, and their composition.

When rocks cool from magma, we call them intrusive igneous rocks. They're called "intrusive" because they form when magma forces its way, or intrudes, into layers of rock in the earth's crust. The rocks form by cooling slowly under the earth's crust. Because of the slow cooling time, large crystals have time to form, organizing layer over layer (consider how rock candy is able to grow large sugar crystals when a string is left in a sugar solution for days). These igneous rocks are also called plutonic rocks, after Pluto, the Roman god of the underworld.

Igneous rocks that cool from lava are called extrusive igneous rocks. A substance extrudes when it oozes out; for example, toothpaste extrudes from the tube when you leave the cap off. When lava oozes out through the earth's crust, we say it "extrudes." Because it's exposed to the air, the lava cools relatively quickly, and the rocks formed from it have small crystals.

Igneous rocks are made of a mixture of naturally occurring minerals. Rocks can vary widely in their composition, but, like a baker's case of cookies and cakes, they all tend to be made from the same ingredients. In baked goods, the proportions of eggs, flour, butter, and so on vary from confection to confection; in different rocks, the proportions of various minerals differ. Some igneous rocks, for example, contain more iron and magnesium than others. These rocks tend to be dense and dark in color. Other igneous rocks contain more silica, the same material that glass is often made from. These rocks tend to be less dense and lighter in color.

The way these rocks form—quickly or slowly, under more or less pressure—and their composition determines their properties: hardness, brittleness, magnetic qualities, color, and so on. These qualities, in turn, determine how we can use the rocks. The composition of rocks also determines whether they will be part of the lower, denser ocean floor or the higher, less dense continental crust.

Under special conditions, rocks containing large amounts of copper, silver, and gold form. We call such rocks metal ores, or ore bodies. Under other conditions, veins containing diamonds and other gemstones crystallize. Conditions that form ore bodies and veins happen only inside the earth's crust. Obsidian, on the other hand, cools very quickly at the earth's surface, which allows no crystals to form. Before the invention of lasers, obsidian scalpels were used in eye surgery because their blades were sharper and harder than steel. ■

READING SELECTION

EXTENDING YOUR KNOWLEDGE

VOLCANOES: ROARING DEMONS RAGING GIANTS

New Zealand, a large island country in the southwestern Pacific Ocean near Australia, is almost always blowing off steam. If volcanoes aren't exploding, then hot springs, geysers, and boiling lakes are fuming. When the British came to explore New Zealand, they found a people called the Maori living there. The Maori like to tell stories. One Maori tale, "How Volcanoes Got Their Fire," tells how fire came to volcanoes in New Zealand. In another tale, "Battle of the Giants," volcanoes act like giant people.

HOW VOLCANOES GOT THEIR FIRE

A powerful medicine man named Ngatoro led his people from Hawaii to New Zealand in canoes. After they arrived, Ngatoro took his female slave, Auruhoe, and climbed the volcano Tongariro. He asked the rest of his people to stop eating until he and Auruhoe returned. He believed this would give him strength against the cold air high on the mountain. Ngatoro and his slave stayed longer than expected. His people got hungry and began eating again. When that happened, Ngatoro and Auruhoe felt the freezing cold. Ngatoro prayed to his sisters back in Hawaii to send fire to warm them. The sisters heard his cry for help and called up fire demons who began to swim under water toward New Zealand. When the fire demons came up at White Island to find out where they were, the earth burst into flames. The demons reached the mainland and continued to travel underground toward Tongariro. Wherever the fire demons surfaced, hot water spewed from the earth and formed a hot spring or geyser. Finally, the fire demons burst out of the top of Tongariro. Their fire warmed Ngatoro and helped save his life, but Auruhoe was already dead from the cold. To remember the journey of Ngatoro and Auruhoe, the Maori called the mountain Ngauruhoe.

HOT WATER VAPOR BLOWS FROM A STEAM VENT IN ROTORUA, NEW ZEALAND. THIS IS JUST ONE OF THE SIGNS OF THE EARTH'S HOT INTERIOR.

PHOTO: Scott Thompson/creativecommons.org

A MAORI KING FROM THE EARLY 1900S

PHOTO: Library of Congress, Prints & Photographs Division, LC-USZ62-109768

READING SELECTION

EXTENDING YOUR KNOWLEDGE

ACCORDING TO THE LEGEND, MT. RUAPEHU GOT ITS BROKEN-UP TOP FROM THE STONES THAT WERE HURLED AT IT BY THE VOLCANO TARANAKI.

PHOTO: Felipe Skroski/creativecommons.org

MT. TONGARIRO IS LOCATED IN TONGARIRO NATIONAL PARK IN NEW ZEALAND.

PHOTO: Strange Ones/creativecommons.org

BATTLE OF THE GIANTS

Three volcanoes—Taranaki, Ruapehu, and Tongariro—lived near each other. Taranaki and Ruapehu both fell in love with Tongariro, but she could not decide which one she preferred. Finally, they decided to fight for her. Tearing himself loose from the earth, Taranaki thrust himself at Ruapehu and tried to crush him. "I'll get you," fumed Ruapehu. He heated the waters in his crater lake until they were boiling. Then he sprayed scalding water over Taranaki and on the countryside around him. The scalding bath hurt Taranaki badly. Furious, he hurled a shower of stones at Ruapehu. The stones broke the top of Ruapehu's cone, which ruined his good looks. "I'll show him," said Ruapehu. He swallowed his broken cone, melted it, and spat it at Taranaki. The molten cone burned Taranaki badly, and he ran to the sea to cool his burns.

For a long time, when Tongariro erupted, the warlike tribes of the area saw it as a sign that they should act like the quarreling giants in the myth, and they made war with each other. Today, the Maori are still afraid that Taranaki will awaken and begin fighting again. They refuse to live or be buried anywhere on a line between Taranaki and the other two peaks. They may be right. Taranaki, 2400 meters (7874 feet) high and snow capped, last erupted only 300 years ago. In their stories, the Maori were probably describing events that they had seen. When a volcano erupts, the top of the mountain sometimes blows off or collapses into the crater. Such an event was probably the basis of the story about Ruapehu's broken cone. Scientists have also shown that the path of lava that flowed from Taranaki to the sea formed New Zealand's Wanganui River Valley. ■

DISCUSSION QUESTIONS

1. Why do people tell stories like "How Volcanoes Got Their Fire"? Are non-scientific stories that explain catastrophic events useful?
2. What sorts of volcanic events might people have been describing in the telling of "Battle of the Giants"?

LESSON 14

VOLCANIC ASH

A HELICOPTER STIRS UP ASH FROM MT. ST. HELENS. ASH FELL IN 11 STATES AFTER THE VOLCANO ERUPTED.

PHOTO: U.S. Geological Survey/Cascades Volcano Observatory/photo by Lyn Topinka

INTRODUCTION

When magma forcefully erupts from a volcano, it can shatter into billions of fragments. These pieces can be as small as dust particles or as large as trucks. The fragmented materials include rocks, minerals, and broken pieces of volcanic glass. When the broken pieces are fine grained, the material is called volcanic ash.

Most people think of a volcanic eruption as fiery streams of red lava and glowing rock. Lava flows can pose great danger to people. They may cover the land and can even set buildings or trees on fire. Lava flows, however, are actually the least dangerous part of a volcanic eruption. Most lava flows do not move particularly fast. And because they are fluid, they tend to flow along low-lying areas. This means that predicting the path of the lava is fairly easy. Although property will be destroyed, areas likely to be affected by the lava can be evacuated.

Ash from violent eruptions, in contrast, often poses great environmental and personal dangers. During an ash fall, the ash that has been ejected into the atmosphere settles back to the earth over a wide area. The ash may cover everything it falls on, smothering crops and coating people's lungs. Or it can erupt into the atmosphere, blocking sunlight, and making daylight turn to darkness. Ash can also move over the ground or close to it in clouds of ash and gas. These clouds, which move like raging rivers, sometimes move at more than 400 kilometers (249 miles) per hour. Melted glaciers, snow, and rain-soaked soils mix with the ash. These mudflows of ash flow down the mountainside, hugging the ground.

In this lesson, you will examine rock fragments that make up volcanic ash and model an ash fall. You will investigate how different-sized volcanic materials erupt into and settle out of the air. You will then discuss how an ash fall can affect the atmosphere, weather, the land, and people and animals.

OBJECTIVES FOR THIS LESSON

- Observe the properties of two igneous rock samples and their fragments.
- Investigate how volcanic particles, such as ash, erupt into and settle out of the air.
- Determine how the size of airborne materials affects where and how fast they settle.
- Draw conclusions about how weather conditions, such as wind, affect the direction and speed at which ash moves.
- Identify the constructive and destructive effects of ash fall.
- Understand that changes caused by volcanic eruptions can occur quickly or slowly over geologic time.

MATERIALS FOR LESSON 14

For you

- 1 copy of Student Sheet 14.1b: Investigating Ash Fall—Observation Sheet
- 1 copy of Student Sheet 14.1c: Volcanoes Review
- 1 copy of the graph grid on Inquiry Master 14.1: Graphing Ash Fall Data
- 1 pair of safety goggles
- 1 dust mask

For your group

- 1 copy of Student Sheet 14.1a: Investigating Ash Fall—Planning Sheet
- 1 plastic box with lid
  - 1 set of rocks, labeled #6 through #9
  - 1 penny
  - 3 hand lenses
  - 1 double-eye loupe
  - 1 sheet of black construction paper
  - 1 sheet of white construction paper (or loose-leaf paper)
  - 1 magnetic compass
  - 1 small plastic bowl
  - 1 metric ruler
  - 1 piece of ash fall paper with double-sided tape
  - 1 bendable straw
  - 1 red marking pen
  - 1 measuring spoon
  - 1 cup of corn kernels, with lid
  - 1 cup of cornmeal, with lid
  - 1 cup of cornstarch, with lid
  - 1 clipboard (optional)
- Rocks, masking tape, or other objects to hold paper down
- Paper towels
- 1 beaker of water

GETTING STARTED

1. Using the materials in your plastic box, examine the rocks labeled #6 and #7. Answer these questions as you observe the rocks with the observation tools.

   A. How are the rocks alike and different?

   B. What are the basic properties of each rock?

   C. Describe how you think the rocks formed. What type of rock is each sample?

   D. Can you identify the rocks by name? Explain.

2. Mark your dust mask with your initials and put it on. Wear safety goggles as well. Working with one rock at a time over black paper, scratch the penny against the rock, as shown in Figure 14.1. Use the loupe and hand lenses to look closely at the rock fragments. Do this with both rocks #6 and #7. Now try white paper. Then answer these questions:

   A. What are the properties of each rock's fragments?

   B. How are the fragments of each rock alike or different?

   C. What do you think causes rocks to break into fine pieces?

3. Use the beaker of water to test the buoyancy of each rock sample. Think about how this property might affect waterways. Dry the rocks with a paper towel when you are finished.

4. Discuss your observations with the class.

5. Now listen as your teacher models and talks about how the rocks you have observed form.

6. The fine-grained rock fragments on your paper are some of the fragments that make up volcanic ash. Brainstorm what you know and want to know about volcanic ash, including how it forms and how it affects people and the environment.

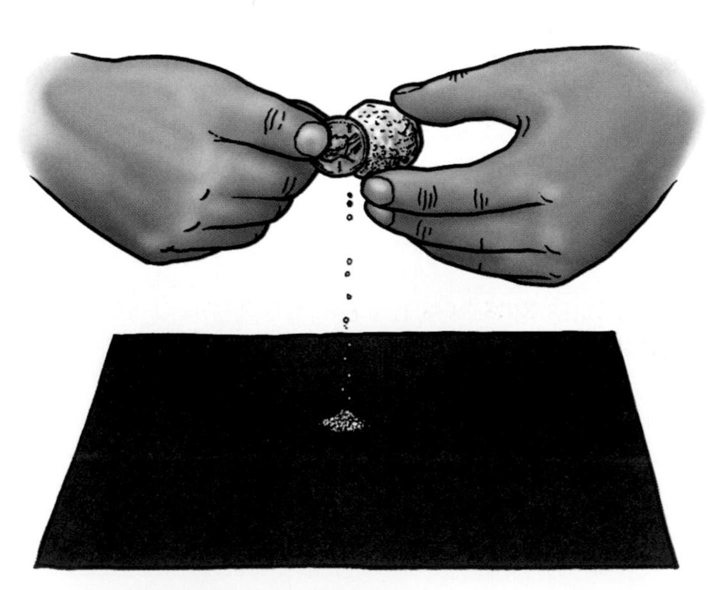

SCRAPE THE PENNY AGAINST EACH ROCK. WHAT DO YOU OBSERVE?
FIGURE **14.1**

INVESTIGATING ASH FALL

PROCEDURE

1. Look at the materials your teacher has set out. How can you use these materials to investigate ash fall? Discuss your ideas with the class, then review with your teacher Procedure Steps 8 through 22.

2. Your teacher will give your group one copy of Student Sheet 14.1a: Investigating Ash Fall—Planning Sheet. What question will your group try to answer based on the suggested procedures you have just discussed? Record that question in the first box on Student Sheet 14.1a.

3. Complete Student Sheet 14.1a by describing the materials and procedures you will use, how many trials you will conduct, what you will measure and how you will measure it, and what you will look for.

4. Discuss your group's plan with your teacher.

5. Discuss with the class how you will record your data and observations and graph your results.

6. Make sure you have your goggles, dust mask, and materials for your group. You will conduct this investigation outside, if possible. If you have a clipboard, put your group's copy of Student Sheet 14.1a on it. Also bring Student Sheet 14.1b: Investigating Ash Fall—Observation Sheet for recording observations.

7. Review the Safety Tips with your teacher. Then perform your investigation by following Procedure Steps 8 through 22.

SAFETY TIPS

Only one member of your group must blow into the bendable straw. Do not share the straw. Throw it out after the class period ends.

Before starting Step 13, everyone in your group needs to put on safety goggles and a dust mask.

The person who will blow through the straw should cut a small hole in his or her dust mask.

Inquiry 14.1 continued

8 Place your construction paper on the ground or work area.

9 Remove the protective paper from the strip of tape. Stick your compass to the tape, as shown in **Figure** 14.2. Turn your paper until your compass matches the direction your group decided on. Record the compass direction on your observation sheet.

PEEL OFF THE PAPER ON THE UPPER SIDE OF THE TAPE. STICK THE MAGNETIC COMPASS TO THE TAPE.
FIGURE **14.2**

10 Place the small bowl in the center of the tape, as shown in Figure 14.3. The bowl will represent a volcano. Tape your paper to the ground, or place rocks or other objects on either end of the paper to hold it down in case the wind blows.

11 Use Student Sheet 14.1a to decide how many corn kernels you will put into the bowl each time. Then fill the bowl with this amount of corn kernels. The corn kernels represents broken pieces of volcanic rock and other large volcanic debris.

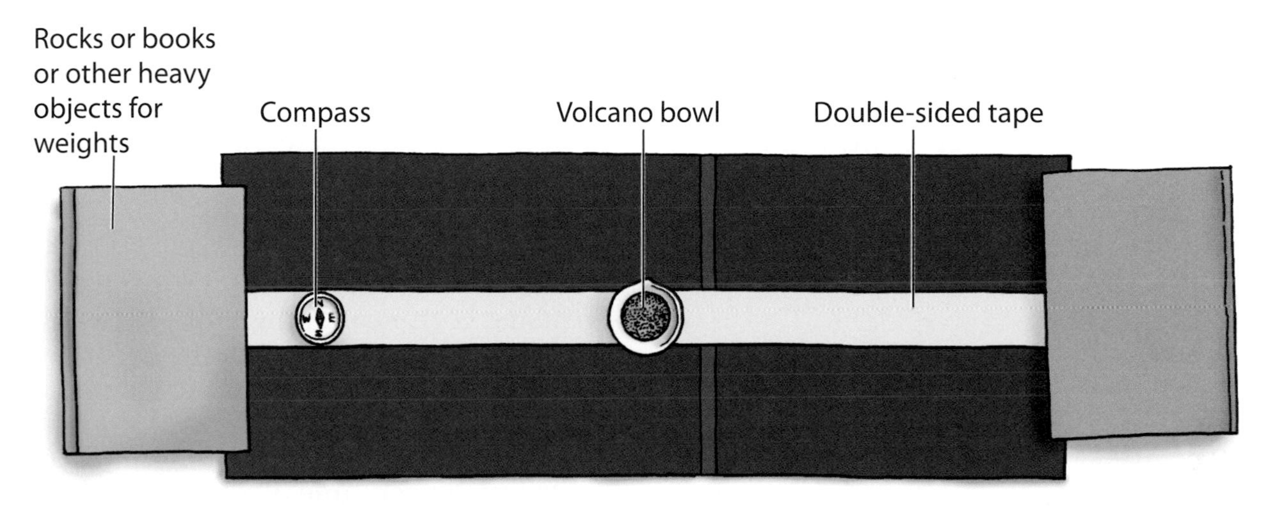

SAMPLE SET UP FOR THE ASH FALL INVESTIGATION.
FIGURE **14.3**

Inquiry 14.1 continued

12. Designate one student to blow into the straw. This will force the gases, ash, and larger particles upward during the model eruption. Make a bend in the straw.

13. The person who blows into the straw should hold it straight and then place the bent end of the straw in the bowl, just touching the surface of the sample, as shown in Figure 14.4. When your group is ready, the student with the straw should blow forcefully into it with *one* quick, sharp blow, while the other members of the group observe how far the particles travel and how long they remain in the air. Watch closely. It happens quickly.

14. Draw your results on Student Sheet 14.1b.

15. Use the red pen to mark on the paper where the corn kernels fell. Draw a line to surround the material, like a fence around horses. You may want to write "Corn Kernels 1" next to it. Then measure how far from the bowl the corn kernels traveled in one direction. Record this distance.

16. Refill the bowl with the same amount of corn kernels and repeat Steps 13 through 15. Do this a total of three times. (Do *not* clean off the kernels between trials. You want them to pile up.) Record all observations, including wind direction and other weather conditions.

17. Repeat Procedure Steps 13 through 15 using cornmeal in the bowl. Remember to use one quick, sharp blow. Conduct three trials, using the same volume of cornmeal each time. Mark your paper and measure how far the cornmeal traveled each time. Record your data and observations.

USE THE STRAW TO FORCE PARTICLES INTO THE AIR.
FIGURE **14.4**

Inquiry 14.1 continued

18. Finally, repeat the steps using cornstarch in the bowl. Use the same volume each time, and measure in the same way each time to keep this a fair test.

19. The particles from a volcanic eruption are a variety of sizes. Create a mixture of the three substances by putting equal amounts of each substance in the bowl. Make a prediction. How will the mixture erupt into the air and settle out around the volcano? What factors will affect how and where the substances settle? Discuss your predictions with your group.

20. Use the straw to blow the mixture into the air. Discuss your results. Were your predictions correct? How was this eruption like a real volcanic ash fall? Add your results to your drawing.

21. Clean up. Throw away the straw. Put your group members' names on the sheet of paper. Keep all extra corn kernels, cornstarch, and cornmeal in the cups. Place a lid on each cup. Other classes will use them.

22. Make certain your data and observation table are complete. Then graph your data on Inquiry Master 14.1: Graphing Ash Fall Data.

REFLECTING ON WHAT YOU'VE DONE

1. Share your graph with the class.

2. As you analyze the graphed data, consider factors that may have affected your results. Answer these questions in your science notebook:

   A. How did the size of the particles affect how far each substance traveled?

   B. How did wind affect the way in which materials erupted into and settled out of the air?

   C. How did the size of the particles affect how long they remained in the air?

3. Apply what you observed in the inquiry to a real eruption of ash and other volcanic fragments. Answer the following:

   A. How might wind or other weather conditions (such as rain) be a factor in a volcanic eruption?

   B. How does hot air behave? Knowing this, why do you think a hot cloud of ash rises so high in the air?

   C. Volcanic ash is a material that absorbs and reflects radiant energy. How might ash clouds affect temperatures on the earth's surface?

   D. If ash repeatedly erupted from a volcano, how might the land change over time?

4. Examine rocks #8 and #9.

   A. How do you think these rocks are related to ash?

   B. How do you think these rocks were formed?

   C. Looking at rocks #8 and #9, do you think there are any constructive effects of ash erupting from a volcano? What might they be?

5. Write a conclusion to your investigation in your science notebook. Be sure to:

   - Answer the question you investigated.
   - Restate your hypothesis and say whether or not it was supported by your data.
   - Include representative data from your investigation to support your answer.

6. Look ahead to Lesson 15, in which your teacher will assess your knowledge of earthquakes, volcanoes, and plate tectonics. To prepare for this assessment, complete Student Sheet 14.1c: Volcanoes Review for homework.

READING SELECTION

EXTENDING YOUR KNOWLEDGE

Mt. St. Helens Erupts

On May 18, 1980, a tremendous volcanic eruption occurred in Washington state. Mt. St. Helens, in the Cascade Range, "blew its top."

Months before the eruption, scientists had observed many signs that the mountain was about to blow. A large bulge on the north face of the mountain kept growing, which was a sign that magma was rising. On March 20, 1980, an earthquake shook the area. It measured 4.1 on the Richter scale. One week later, a series of explosions began that sent fragmented older volcanic rock and steam into the air. These earthquakes, together with periodic venting of rock and steam, continued for weeks. Sulfuric acid levels rose in local ponds and streams, and hydrogen sulfide odors increased dramatically.

On May 18, 1980, an earthquake that registered 5.0 on the Richter scale triggered the collapse of the bulging north side of the mountain, causing a volcanic landslide. The decrease in pressure on the magma chamber caused a violent release of steam and lava. As bubbling lava rose in the air, it solidified instantly. A fine ash cloud rose 19 kilometers (12 miles) above the volcano, as shown in the photo at right.

Ash fell several meters deep in areas close to Mt. St. Helens, while prevailing winds drove the cloud to the east-northeast. Communities as far away as 800 kilometers (about 500 miles) were blanketed by ash. In Yakima, Washington, 130 kilometers (81 miles) to the east, the ash fall caused almost total darkness at midday. Nearly 1 billion tons of ash were deposited over a huge area. Acid droplets from the eruption remained suspended in the atmosphere for as long as two years.

MT. ST. HELENS IS SHOWN HERE BEFORE ITS ERUPTION ON MAY 18, 1980.

PHOTO: U.S. Geological Survey/Cascades Volcano Observatory/photo by Rick Hoblitt

WHEN MT. ST. HELENS ERUPTED, AN ASH CLOUD ROSE 19 KILOMETERS (12 MILES) ABOVE THE VOLCANO.

PHOTO: U.S. Geological Survey/Cascades Volcano Observatory/ photo by Mike Doukas

Flows of hot gases and volcanic ash more dense than air raced down the north side of the mountain. This heavy ash flow destroyed everything in its path. It caused steam explosions when it encountered bodies of water or moist ground. These explosions continued for weeks; one even occurred a year later.

Heat from the eruption melted snow and glaciers, which mixed with ash on the upper slopes of the mountain and formed a thick volcanic mudflow. Like an avalanche, the mudflow displaced water in lakes and streams and caused flooding downstream.

The volcano's warning signs had allowed scientists to warn government agencies, which closed much of the area to tourists and restricted the activity of residents. Sadly, however, 63 people died as a result of the eruption, most of them because they ignored the evacuation advice.

READING SELECTION

EXTENDING YOUR KNOWLEDGE

The eruption of Mt. St. Helens left a large crater. Five more explosive eruptions occurred during 1980, and the volcano continued to erupt through 1986. These successive eruptions created a lava dome on the floor of the crater. Today, the eruptions appear to be over. But Mt. St. Helens is the most frequently active volcano in the Cascade Range, and scientists anticipate the volcano will erupt violently again.

MONITORING THE VOLCANO'S WARNING SIGNS

Scientists at the U.S. Geological Survey monitor Mt. St. Helens in order to predict future eruptions. In the photo on page 198, geologists use a steel tape to measure the distance across a crack in the volcano's crater floor. Widening cracks indicate that magma is rising, deforming the area, and leading to an eruption. These cracks usually extend outward from the lava dome, like the spokes of a wheel.

DURING THE MAY 1980 ERUPTION OF MT. ST. HELENS, AT LEAST 17 SEPARATE ASH FLOWS MOVED DOWN THE MOUNTAIN LIKE RAGING RIVERS. THEY TRAVELED AT SPEEDS OF OVER 100 KILOMETERS (62 MILES) PER HOUR AND REACHED TEMPERATURES OF OVER 400°C (752°F).

PHOTO: U.S. Geological Survey/Cascades Volcano Observatory/photo by Peter W. Lipman

PLUMES OF STEAM, GAS, AND ASH, WHICH ROSE 1000 METERS (ABOUT 3280 FEET) INTO THE AIR, ERUPTED FROM THE CRATER. SOME PLUMES WERE SEEN FROM PORTLAND, OREGON, 81 KILOMETERS (50 MILES) TO THE SOUTH.

PHOTO: U.S. Geological Survey/Cascades Volcano Observatory/photo by Lyn Topinka

NEARLY 220 KILOMETERS (137 MILES) OF RIVER CHANNELS SURROUNDING THE VOLCANO WERE AFFECTED BY MUDFLOWS. A MUD LINE ON THE TREES SHOWS THE DEPTH OF THE MUD.

PHOTO: U.S. Geological Survey/Cascades Volcano Observatory/photo by Lyn Topinka

Geologists also use a tiltmeter to electronically measure changes in the slope of the crater floor, which are caused by moving magma. Tiltmeters allow 24-hour monitoring. The information collected from these instruments is relayed to the volcano observatory.

Scientists also place seismographs at stations near the lava dome to monitor earthquake activity. An increase in the number of earthquake vibrations is often the first sign that a major eruption is approaching. Scientists also collect gas samples from the volcano. They place gas sensors around vents near the lava dome and crater floor. Specially equipped airplanes measure sulfur dioxide gas, which usually increases 5- to 10-fold during an eruption.

SIGNS OF RENEWAL

Plant and animal life returned to Mt. St. Helens. As early as the summer of 1980, new vegetation began to appear. Many small trees and plants, which had been protected during the eruption by packed snow, re-emerged after snowmelt. Seeds carried by the wind or by animals landed in the area and sprouted on the lava-covered ground. By 1985, new growth covered all the ridges surrounding the volcano. During the May 1980 eruption, many small animals—such as gophers, mice, frogs, fish, and insects—were protected from the blast because they were below ground or underwater. Many large animals, such as bear, elk, deer, and coyotes, were killed in the eruption, but with the return of a food supply, they have repopulated the region.

READING SELECTION

EXTENDING YOUR KNOWLEDGE

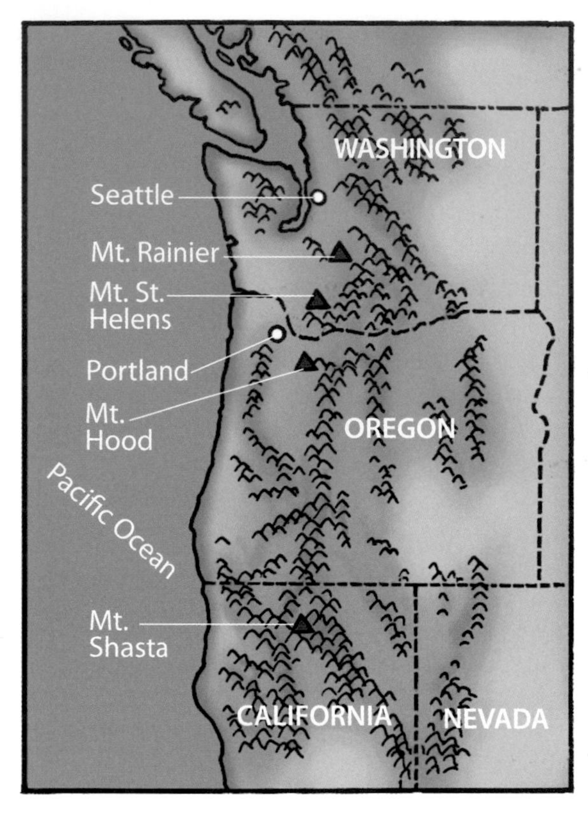

MAP OF THE CASCADE RANGE

FUTURE ERUPTIONS?

Several other mountains in the Cascade Range (see the map at left) pose a threat to populated areas. Eruptions of Mt. Shasta in northern California would cause damage and perhaps fatalities to several nearby communities. Mt. Hood, in Oregon, lies less than 65 kilometers (40 miles) away from the densely populated city of Portland. Probably the most dangerous eruption would be from Mt. Rainier, in Washington. During the past 10,000 years, there have been at least 60 mudflows from Mt. Rainier, one of which completely covered an area that is now populated by 120,000 people. No one can predict when another mudflow from Mt. Rainier might take place or when any of these Cascade mountain volcanoes might "wake up." With continued monitoring and emergency evacuation plans in place, scientists and public officials hope everyone will be ready when the next volcano blows its top. ■

SCIENTISTS USE A STEEL TAPE TO MEASURE THE CHANGES IN THE CRACKS ON THE CRATER FLOOR.

PHOTO: U.S. Geological Survey/Cascades Volcano Observatory/photo by Lyn Topinka

GEOLOGISTS COLLECT GAS SAMPLES AROUND THE DOME OF THE VOLCANO.

PHOTO: U.S. Geological Survey/Cascades Volcano Observatory/photo by Thomas J. Casadevall

FIREWEED IS ONE SPECIES OF PLANT THAT RETURNED TO MT. ST. HELENS.

PHOTO: U.S. Geological Survey/Cascades Volcano Observatory/photo by Lyn Topinka

DISCUSSION QUESTIONS

1. What characteristics made the 1980 eruption of Mt. St. Helens particularly dangerous to people and wildlife?
2. What allowed most people to escape its destructive effects?

READING SELECTION

EXTENDING YOUR KNOWLEDGE

VOLCANO IN A CORNFIELD

In the mid-1900s, a farmer named Dionisio Pulido lived with his family in the small town of Paricutín, Mexico. Pulido's farm was a few miles outside the town. He had owned it for 31 years. There was nothing unusual about this farm, except for a small depression in the cornfield. Pulido and his wife had tried to fill the depression several times, but it always came back. In fact, one local resident recalled that, 50 years earlier as a small child, he had played near the "small hole." He remembered hearing underground noises like falling rocks near the hole and felt "a pleasant warmth" coming from it.

The year 1943 started off as usual in Paricutín. There were reports of small earthquakes in the area, but no one was too concerned. Earthquakes were common in these parts. So, on February 20, Pulido mounted his horse and set off as usual to prepare his fields for spring planting. Then something very unusual happened.

"I heard a noise like thunder during a rainstorm, but I could not explain it, for the sky above was clear and the day was so peaceful," he recalled.

Pulido soon noticed something else in his field: along with the depression, a crevice had opened up on the side of a hill. "Here is something new and strange," he remembered thinking.

THE VOLCANO IN PULIDO'S FIELD WAS LATER NAMED PARICUTÍN. HERE IS THE VOLCANO SHORTLY AFTER IT FORMED IN 1943.

PHOTO: K. Segerstrom, U.S. Geological Survey/National Geophysical Data Center/NOAA

THE VOLCANO PARICUTÍN GREW QUICKLY.

PHOTO: U.S. Geological Survey

READING SELECTION

EXTENDING YOUR KNOWLEDGE

ASH FROM PARICUTÍN COVERED THE TOWN AND SURROUNDING FIELDS.

PHOTO: U.S. Geological Survey

A few minutes later, he heard thunder again, and he saw the trees tremble. "It was then I saw how, in the hole, the ground swelled and raised itself up 2 or 2.5 meters (6.6 or 8.2 feet) high, and a kind of fine dust—gray, like ashes—began to rise up in a portion of the crack. Immediately, more clouds of dust began to rise, with a hiss or whistle, loud and continuous. A smell of sulfur filled the air. I became greatly frightened."

When sparks ignited pine trees 23 meters (75 feet) from the jagged rip in the ground, Pulido raced back to his home.

Throughout the night, the volcano continued to grow. By midnight, huge incandescent bombs were hurling into the air with a roar. Lightning flashes appeared in the heavy columns of ash. In the morning when Pulido returned to his cornfield, he saw an amazing sight. A cinder cone—almost 2 meters (6.6 feet) high—had formed where the depression and crevice had been. The cone was spitting out fumes, clouds of ash, and rocks with great violence. By midday, the cone was nearly 10 meters (33 feet) high and still growing. That night, a neighbor described the scene: "Stones rose to a height of 500 meters (1640 feet). They flew through the air to fall 300 to 400 meters (984 to 1312 feet) from the vent … on the plowed fields where I used to watch the cattle of my grandfather."

On the night of February 22, Ezequiel Ordonez, a Mexican geologist, arrived on the scene and recorded it officially: "I was witnessing a sight which few other humans had ever seen, the initial stages of the growth of a new volcano!"

On the second day, the hill had grown to 30 meters (98 feet), and by the third day, it had grown to 60 meters (197 feet). On the sixth day, it was 120 meters (394 feet). By the end of one month, the cinder cone was 148 meters (486 feet) high.

But the volcano's size was not the real problem. The problem was ash and lava that poured out of it.

Lava flows broke through the sides and base of the cone. They deposited a mass of black, jagged rocks about 1 meter (3.3 feet) deep all over Pulido's farm. Then the lava began to advance toward the town. By June, the situation became desperate. Ash soon covered the town of Paricutín and the surrounding farms. Thousands of cattle and hundreds of horses died from breathing the ash. When the seasonal rains resumed in May, the ash turned to muddy rain. The farms were ruined. Government officials

A FLOW OF BLOCK LAVA FROM PARICUTÍN IN 1994, WITH ASH-COVERED FIELDS IN THE FOREGROUND

PHOTO: U.S. Geological Survey

and geologists agreed that the town had to be evacuated. Some residents waited until the lava was at their back door before reluctantly leaving. Pulido and his family were among those who had to leave their farms.

By the end of September, the town of Paricutín had disappeared. The river of molten rock and ash had set fire to it. Today, the ruins of Paricutín and much of the nearby countryside still lie buried beneath lava rock. ■

DISCUSSION QUESTIONS

1. Why do you think the farmers around Paricutín were so reluctant to leave the site of the eruption?
2. If you had to figure out where a new volcano is likely to form, what sorts of things would you look for?

ASH GOES INTERNATIONAL

On April 14, 2010, mass confusion reigned at airports in London. Travelers managing luggage, business equipment, and children arrived at the airports only to find guards and airline staff directing thousands of would-be passengers away from the gates, away from the planes. A newspaper headline explained: "ALL FLIGHTS CANCELLED DUE TO ASH CLOUD FROM EYJAFJALLAJÖKULL."

It sounded like a joke, but it wasn't funny to stranded travelers who wondered how they would find their ways back home. Eyjafjallajökull (pronounced EY-ya-fyat-lah-YOH-kuht, or, according to *The New York Times*, "Hey, ya fergot La Yogurt") was a fissure in Iceland. Even as the world struggled to pronounce its name, it was perfectly clear that grounded travelers were in trouble. Authorities repeated the message: There was no knowing how long the airports might be closed.

THE ERUPTION OF EYJAFJALLAJÖKULL

PHOTO: NASA image by Jeff Schmaltz, MODIS Rapid Response Team

NATO PLANES LIKE THESE WERE USED TO DETERMINE THE EFFECTS OF THE ASH CLOUD ON AIRPLANE ENGINES.

PHOTO: DoD photo by Senior Airman Greg L. Davis, U.S. Air Force

A PLUME OF ASH FROM THE EYJAFJALLAJÖKULL ERUPTION MOVES SOUTHEAST OVER THE NORTH ATLANTIC OCEAN.

PHOTO: NASA image by Jeff Schmaltz, MODIS Rapid Response Team at NASA GSFC

The eruption of Eyjafjallajökull actually began in March 2010, with fire fountains and lava flows, but not much ash. Then, with little warning, on April 14, new craters opened in the volcano and a giant ash plume of steam, fine silica particles, and pulverized rock shot skyward. Carried by the jet stream, a fast moving current of air about 9000 meters (about 30,000 feet) above Earth's surface, the ash plume reached the airspace over mainland Europe within a day.

Airlines reacted quickly, cancelling all flights in and out of most European countries, including England, Scotland, Ireland, and France. While it was hard for scientists to determine how dangerous the material in the ash plume was, previous experience suggested that it might cripple airplane engines. In 1989, a plane had lost power in its engines when it flew through a cloud from the eruption of Mt. Redoubt in Alaska. Engineers feared that particles of silica—a glasslike material found in sand, also known as silica dioxide—might be sucked into the hot engine, melt to form a glassy coating over engine parts and ventilation holes, and cause the engine to overheat or otherwise fail. The ash might also have damaged airplane windshields, making it tough for pilots to see out.

What made the volcano suddenly erupt so violently? The eruptive areas lay beneath a glacier on Mt. Eyjafjallajökull, and the combination of the extremely hot magma and glacial ice proved explosive. Cold meltwater ran down into the volcanic vent and hit the magma, whereupon the volcano erupted explosively,

READING SELECTION

EXTENDING YOUR KNOWLEDGE

with steam shooting up and carrying liquid magma. But the glacial ice and meltwater also cooled the magma quickly, turning it into tiny, hard fragments. The eruption's heat also vaporized some ice, generating a powerful column of steam, which carried the ash (made of fragments of rock and silica) as high as 7300 meters (about 24,000 feet). The ash was then given a ride across the North Atlantic on the jet stream, which travels south and east from Iceland and over the European continent.

As the ash spread over Europe, thousands of passengers were stranded. Ships, buses, and trains filled up with people willing to pay extra to get home; they were missing important events, such as weddings, funerals, and the start of school. As no air freight could travel, worries mounted about how to keep people supplied with food and medicines in Europe.

The North Atlantic Treaty Organization (NATO), an international military alliance, was asked to help. NATO sent some F-16 fighter jets through the ash cloud to see whether it might be possible for airlines to fly safely out. One of the plane's engines suffered glassy deposits from the fly-through, confirming that no commercial planes should fly until the ash cloud moved on.

But Europeans and travelers were not the only victims of Eyjafjallajökull. Consider the Icelanders themselves. Many lived on farms near Eyjafjallajökull, which were suddenly flooded as meltwater poured off the volcano. Thanks to evacuations, no Icelanders were known to have died in the flooding, but people returned to find their farms waterlogged and covered in ash.

Despite its powerful effects, the 2010 eruption of Eyjafjallajökull was mild compared to an eruption that occurred in Iceland a few centuries ago. On June 8, 1783, the crater of a fissure volcano called Laki burst open, freeing lakes of boiling lava. For nearly eight months lava spilled out, covering about 600 square kilometers (232 square miles) of countryside and killing about 10,000 people. A cloud of gases containing hydrofluoric acid and sulfur dioxide spewed out of the volcano and was carried over Iceland and to Europe on the winds. Unfortunately, sulfur dioxide reacts with water—even the water in human lung tissue—to form poisonous sulfuric acid. Tens of thousands of Europeans choked to death from inhaling the gases. A terrible winter followed, as Icelanders and their livestock died of sulfur and fluorine poisoning and famine. About a quarter of the population died. The winter was unusually hot; a lingering haze of ash and sulfur dioxide hung over Iceland and western Europe. Chroniclers of the time wrote that fresh-slaughtered meat had to be eaten immediately, because it would go bad by the next day.

Indeed, Iceland, sometimes called the land of fire and ice, is a highly volcanic place. It is part of the Mid-Atlantic Ridge (which you read about in Lesson 6), an area where magma is rising and plates are spreading. What is unusual about Iceland is that the ridge is above sea level, which means that violent eruptions are not buffered by the ocean.

There is concern that the 2010 eruption of Eyjafjallajökull forecasts a worse eruption of the nearby Katla volcano. Every time in history that Eyjafjallajökull has erupted, Katla has followed. Katla is larger, with a magma chamber ten times the size of Eyjafjallajökull's. And because Katla is beneath a huge glacier, an eruption could produce heavy flooding and an avalanche of giant chunks of ice. Scientists have been keeping a close eye on Katla for signs that it may erupt. Monitoring the effects of the Eyjafjallajökull eruption has allowed scientists to learn more about glaciovolcanoes (ice-covered volcanoes). While the volcanoes themselves are similar to other volcanoes, their interactions with the ice when they erupt make them unique. ■

PASSERSBY HAVE USED LAVA ROCKS TO BUILD HUNDREDS OF "CAIRNS" (LANDMARKS MADE OF PILED ROCKS) ON A FARM IN ICELAND.

PHOTO: :mrMark:/creativecommons.org

DISCUSSION QUESTIONS

1. Why did a volcanic eruption in Iceland close airports in Europe but not in North America?
2. Explain the statement "Every volcano is unique" and how this impacts our ability to predict and respond to volcanic eruptions.

LESSON 15

EXPLORING PLATE TECTONICS ASSESSMENT

IN PART A OF YOUR ASSESSMENT, YOU WILL SET UP AN INVESTIGATION TO COMPARE THE EFFECT OF LOOSE SOIL AND PACKED SOIL ON THE WAY MODEL BUILDINGS RESPOND TO SHAKING.

PHOTO: © Terry G. McCrea/ Smithsonian Institution

INTRODUCTION

You have now finished your investigations of plate tectonics. During the next four periods, you will complete a two-part assessment. On the first day, you will review what you have learned in this unit. On the second day, you will complete Part A of the assessment. You will design and carry out an investigation to test the effects of soils on building stability during an earthquake. Working individually, you will plan an investigation, record a plan, state the hypothesis, conduct the investigation, and record observations as they relate to the concepts and skills addressed in *Exploring Plate Tectonics.* On the third day, you will complete Part B, which is a set of multiple-choice and short-answer questions. On the fourth day, you will review the assessment and personally assess how well you have learned the concepts and skills in this unit.

OBJECTIVES FOR THIS LESSON

- Review and reinforce concepts and skills from *Exploring Plate Tectonics.*
- Use knowledge and data interpretation skills to answer questions.
- Design and conduct an experiment to investigate the effect of loose soil versus packed soil on the way model buildings respond to shaking.

MATERIALS FOR LESSON 15

MATERIALS FOR PART A: *EXPLORING PLATE TECTONICS* PERFORMANCE-BASED ASSESSMENT

For you

1 copy of Student Sheet 15.1a: *Exploring Plate Tectonics* Performance-Based Assessment— Planning and Observation Sheet

For you and your partner

1 copy of Inquiry Master 15.1a: *Exploring Plate Tectonics* Performance-Based Assessment
2 medium-sized plastic cups
1 large plastic cup of sand, with lid
1 small cup of diluted glue, with lid
6 pennies
1 measuring spoon
Paper towels

GETTING STARTED

1. Work as directed by your teacher and review what you have learned in this unit. Discuss your responses to Student Sheet 14.1c: Volcanoes Review with the class or with your group. You may also want to review your responses to Student Sheet 8.1b: Earthquakes Review, which you completed earlier in the unit.

PART A

EXPLORING PLATE TECTONICS PERFORMANCE-BASED ASSESSMENT

PROCEDURE

1. Your teacher will give you your own copy of Student Sheet 15.1a: *Exploring Plate Tectonics* Performance-Based Assessment—Planning and Observation Sheet. You will fill out the first page of this planning sheet before starting the investigation. You will record your observations and conclusions on the other pages.

2. Your teacher will show you one set of materials. Although you will share materials with a partner, you are to complete the written portion of the assessment by yourself, unless your teacher tells you otherwise.

3. Discuss with your teacher how you will be assessed in this part of the lesson.

4. Collect your materials. Check them against the materials list to make certain you have everything before you begin.

5. Complete Part A of the assessment. Give Inquiry Master 15.1a and Student Sheet 15.1a to your teacher when you are finished.

6. Clean up by doing the following:

   A. Pour the dry sand back into the larger plastic cup.

   B. Pour your sand-and-glue mixture into the large container your teacher has set out.

   C. Top off the large plastic cup with new dry sand from the distribution center.

   D. Refill your small cup with diluted glue. Replace the lid.

   E. Wipe off the pennies and empty plastic cups with the paper towels.

PART B

EXPLORING PLATE TECTONICS WRITTEN ASSESSMENT

PROCEDURE

MATERIALS FOR PART B: *EXPLORING PLATE TECTONICS* WRITTEN ASSESSMENT

For your group

- 1 copy of Inquiry Master 15.1b: *Exploring Plate Tectonics* Written Assessment
- 1 copy of Student Sheet 15.1b: *Exploring Plate Tectonics* Written Assessment Answer Sheet (or other answer sheet)

1. Your teacher will give you one copy of Inquiry Master 15.1b: *Exploring Plate Tectonics* Written Assessment and one copy of Student Sheet 15.1b: *Exploring Plate Tectonics* Written Assessment Answer Sheet. Write all your answers on the student sheet. Do *not* write on the inquiry master.

2. Complete Part B of the assessment.

3. When you are finished, give your teacher your copies of Student Sheet 15.1b and Inquiry Master 15.1b.

REFLECTING ON WHAT YOU'VE DONE

1. Discuss the experimental designs that the class used to complete the performance assessments, and go over the correct answers to the multiple-choice and short-answer questions. Ask questions about any answers you do not understand.

2. Review your concept map and brainstorming lists from Lesson 1. Add new ideas or anything that you have learned in the unit. Do you still have questions about earthquakes, volcanoes, and plate tectonics? If so, add them to the list.

Glossary

aftershock: An earthquake wave that follows the main shock of an earthquake.

ash: Fragmented volcanic material with particles that measure less than 2 mm in diameter.

ash fall: The ejection of volcanic materials into the atmosphere and the settling of these materials over a wide area of the surface of the earth.

asthenosphere: The layer of the mantle that lies directly below the lithosphere and flows, like taffy.

body wave: An earthquake wave that travels through the body of the earth rather than on its surface. See *P-wave; S-wave.* See also *surface wave.*

brittle: Describes objects that break easily when a force is applied to them. See also *ductile.*

caldera: A large, steep-sided, circular or oval volcanic depression that forms when magma retreats or erupts from a shallow underground magma chamber; with no magma to support the ground above it, the overlying rock collapses and the caldera is formed.

cinder cone: A small (less than 400 meters high), cone-shaped volcano made of broken rocks or blobs of hardened lava, called "cinders," that accumulate around the volcanic vent. See also *composite volcano; shield volcano.*

composite volcano: A volcano that forms from alternating eruptions of viscous lava and broken rock. It is steep near the summit and flat toward the base. See also *cinder cone; shield volcano.*

constructive: Building up, beneficial.

convection: The process by which heat moves efficiently through air or water.

convection cell: A circulating flow of air or water resulting from temperature differences; also called a convection current.

core: The earth's innermost layers, consisting of a liquid iron outer core and a solid iron-nickel inner core. See also *crust; mantle.*

crater: A small, bowl-shaped hole that forms when rock explodes from a volcano during an eruption.

crust: The earth's outer layer; the coolest and least dense layer of the earth. See also *core; mantle.*

deep-sea trench: A deep, narrow depression in the seafloor.

destructive: Causing damage or injury.

ductile: Describes objects that bend, stretch, or flow when a force is applied to them. See also *brittle.*

earthquake: Vibrations in the earth caused by the sudden release of energy, usually as a result of the movement of rocks along a fault.

epicenter: The point on the surface of the earth directly above the focus of an earthquake.

extrusive igneous rock: Igneous rock formed by lava cooling on the surface of the earth.

fault: A fracture in bedrock, along which blocks of rock on opposite sides of the fracture move.

focus: The location where the rupture of an earthquake begins and energy is released.

force: A push or pull on a object.

friction: A force that opposes the motion of objects.

geologist: A scientist who studies the history and structure of the earth as it is recorded in rocks.

globe: A spherical model of the earth.

hot spot: A localized zone of melting in the mantle that is fixed under a plate.

igneous rock: A solid earth material that forms when magma or lava cools and crystallizes on or below the earth's surface. See also *metamorphic rock; sedimentary rock.*

intensity: A measure of the damage done by an earthquake. Determined on the basis of the earthquake's effect on people, structures, and the natural environment.

intrusive igneous rock: Igneous rock that forms from magma cooling inside the earth.

jet stream: A long, narrow current of very strong winds in the upper troposphere.

lag time: The time between the arrival of P-waves and S-waves at a location where an earthquake occurs.

lahar: Mudflow that occurs when rain falls through clouds of ash or when rivers become choked with falling volcanic debris.

landform: A physical feature of the earth's surface, such as a mid-ocean ridge, a trench, or a mountain.

lava: Magma that has reached the surface of the earth.

lava dome: A bulbous, steep-sided dome that forms at the top of a volcano when thick, relatively "cold" magma emerges from the volcanic opening.

lava flow: Lava that flows quickly over the surface of the earth and covers a wide area.

lithosphere: The cool, solid outer shell of the earth. It consists of the crust and the rigid uppermost part of the mantle and is broken up into segments, or plates.

magma: Hot, molten rock inside the earth.

magnitude: A measure of the total amount of energy released at the focus of an earthquake.

mantle: The layer of the earth beneath the crust. It is about 2900 km thick, and it makes up about 83 percent of the earth's interior. See also *core; crust.*

map: A representation of the earth or a part of the earth, usually on a flat surface.

metamorphic rock: A solid earth material that forms when any rock type is changed by the earth's high temperature and pressure. See also *igneous rock; sedimentary rock.*

Mid-Atlantic Ridge: A zone of intense earthquake and volcanic activity that runs down the middle of the Atlantic Ocean floor.

mid-ocean ridge: A mountainlike landform that develops when plates separate and new ocean lithosphere forms.

mineral: A naturally formed, inorganic solid composed of one or more elements.

mitigate: To make something less severe, intense, or painful.

model: A representation that is used to study objects, ideas, or systems that are too complex, distant, large, or small to study easily firsthand.

mudflow: A powerful "river" of mud that forms when debris, such as ash from a volcanic eruption, moves into a stream or river.

natural catastrophic event: A powerful and often dramatic force of nature that changes the earth's surface and atmosphere; includes earthquakes, volcanoes, and intense storms such as hurricanes and tornadoes.

newton: Unit of force in the metric system.

pillow lava: Balloonlike mounds that form when lava flows under water or into the ocean.

plate: A large, mobile segment of the earth's lithosphere.

plate boundary: A place where pieces of the broken lithosphere meet. Boundary types include convergent, divergent, and transform.

plate tectonics: A theory that the lithosphere is broken into segments, or plates, that "float" on the asthenosphere, and that interactions among these plates are associated with earthquakes and volcanic activity and form mid-ocean ridges, trenches, mountains, and chains of volcanic islands.

P-wave: A primary (compressional) earthquake wave that travels through the body of the earth; so named because it is the first wave to reach a seismograph station during an earthquake. See also *S-wave.*

pyroclastic: Made of hot fragments from volcanic material.

Ring of Fire: A zone of intense earthquake and volcanic activity that encircles the Pacific Ocean basin; also called the Circum-Pacific Belt.

risk: Exposure to the chance of injury or loss.

rock: A solid earth material made of various minerals. Igneous, metamorphic, and sedimentary are three types of rock.

rock cycle: The process by which earth materials transform from one rock type into another.

sedimentary rock: A solid earth material composed of compacted and cemented sediments or of particles of various sizes. See also *igneous rock; metamorphic rock.*

seismic wave: A wave generated by earthquake vibrations. See also *P-wave; S-wave; surface wave.*

seismogram: The record made by a seismograph; the paper on which earthquake waves are recorded.

seismograph: An instrument that detects, records, and measures the vibrations produced by an earthquake.

seismologist: A scientist who studies earthquakes.

seismology: The study of earthquakes.

shield volcano: A wide, broadly sloping volcano that forms from runny lava. It is the largest of volcanoes. See also *cinder cone; composite volcano.*

subduction: The movement of a tectonic plate beneath another plate.

surface wave: An earthquake wave that travels on or near the surface of the earth.

S-wave: A secondary earthquake wave; so named because it travels slower than a primary wave and is the second wave to reach the seismograph station after an earthquake. It travels through the body of the earth as a series of crests and troughs. See also *P-wave.*

temperature: A measure of how hot or cold a material is; an indication of the amount of heat energy that has been absorbed by the material.

trench: A deep gorge formed on the ocean floor when an oceanic plate moves beneath a continental plate.

travel time: The time it takes for a wave to travel from one point to another; the time it takes for an earthquake wave to travel from the epicenter of a quake to a location on the earth.

tsunami: A series of sea waves caused by underwater earthquakes or, more rarely, by volcanoes.

upwelling: The rising of cold, deep water from an ocean bottom.

viscosity: The tendency to resist flowing; a property of liquids.

volcano: A landform, usually cone-shaped, produced by a collection of erupted material around a vent, or opening, on the surface of the earth and through which gas and erupted material pass.

volcanologist: A scientist who studies volcanoes and volcanic phenomena.

Index

Photo Credits

Front Cover
NASA image courtesy Jeff Schmaltz, MODIS Land Rapid Response Team at NASA GSFC

Lessons
2 U.S. Geological Survey/photo by Walter D. Mooney, Ph.D. **4** Kendra Helmer/U.S. Agency for International Development (USAID) **6** longhorndave/creativecommons.org **7** McKay Savage/creativecommons.org **8** U.S. Geological Survey/photo by Walter D. Mooney, Ph.D. **10** Phil Whitehouse/creativecommons.org **12** © 2011 Photos.com, a division of Getty Images. All rights reserved. **20** U.S. Geological Survey/photo by S.D. Ellen **22** U.S. Geological Survey/photo by Walter D. Mooney, Ph.D. **23** U.S. Air Force photo by Master Sgt. Jeremy Lock **24** U.S. Geological Survey/photo by Walter D. Mooney, Ph.D. **26** U.S. Geological Survey/photo by Sue Hough **34** NOAA/National Geophysical Data Center **45** FEMA/Jocelyn Augustino **46 (top)** FEMA/Marvin Nauman **(bottom)** FEMA/Jocelyn Augustino **47 (top)** FEMA/Andrea Booher **(bottom)** FEMA/Jocelyn Augustino **48** U.S. Department of Defense **50** NASA Goddard Flight Center **56** U.S. Geological Survey/photo by Walter D. Mooney, Ph.D. **57** National Earthquake Information Center, U.S. Geological Survey, Denver, CO **59** U.S. Geological Survey/photo by Sarah C. Behan **61 (top right)** Library of Congress, Prints & Photographs Division, LC-USZ62-17359 **(bottom left)** Library of Congress, Prints & Photographs Division, LC-USZ62-64748 **63** Stefan Lins/creativecommons.org **64 (left)** Earthquake Hazards Program/U.S. Geological Survey **(right)** Kyle Nishioka, kylenishioka.com/creativecommons.org **65** Steven Brener/creativecommons.org **66** National Cancer Institute/Linda Bartlett **68** Scott Bauer, Agricultural Research Service/U.S. Department of Agriculture **72** National Ocean Service/NOAA **73** National Ocean Service/NOAA **74** Anne Williams/©NSRC **76** NASA image courtesy Jeff Schmaltz, MODIS Rapid Response Team at NASA GSFC **84** Astronaut photograph ISS008-E-13304 taken from the International Space Station on January 28, 2004. Image provided by the Earth Observations Laboratory, Johnson Space Center. **86** Courtesy of Smithsonian Institution Libraries, Dibner Library of the History of Science and Technology, Washington, DC **90** Naotake Murayama/creativecommons.org **92** U.S. Geological Survey **94** R.E. Wallace/USGS/NGDC/NOAA, Boulder, CO **96** U.S. Geological Survey **97** U.S. Geological Survey/photo by John Pallister **101** Sue Hirschfeld/National Geophysical Data Center/Noaa, Boulder, CO **102** Courtesy of Smithsonian Institution Libraries, Dibner Library of the History of Science and Technology, Washington, DC **104** Randolph Femmer/life.nbii.gov **108** Dave Glass/creativecommons.org **109 (top)** NOAA Central Library **(bottom)** Archival Photography by Steve Nicklas, NOS, NGS/NOAA **110** LWY/creativecommons.org **111** NASA/JPL/Malin Space Science Systems **112** NPS photo by Jim Peaco **114** NASA image created by Jesse Allen, using EO-1 ALI data provided courtesy of the NASA EO-1 Team **116** U.S. Geological Survey/

Cascades Volcano Observatory/photo by Peter Lipman **117** U.S. Geological Survey/photo by J.D. Griggs **118** U.S. Geological Survey/Cascades Volcano Observatory/photo by Lyn Topinka **119 (top)** NPS photo by Frank Balthis **(bottom)** P. Hedervari, National Geophysical Data Center/NOAA **120** NPS Photo **122 (top)** NPS photo by J.R. Douglass **(bottom)** NPS photo by George Marle **124** Patrick M. Bonafede/U.S. Agency for International Development (USAID) **126** FEMA/Adam Dubrowa **131** © Terry G. McCrea/Smithsonian Institution **132** © Terry G. McCrea/Smithsonian Institution **135** U.S. Navy photo by Photographer's Mate Airman Jordon R. Beesley **136 (top)** NOAA **(bottom)** NOAA **137** U.S. Geological Survey **138** NASA Jet Propulsion Laboratory **140** OAR/National Undersea Research Program (NURP) **142** U.S. Geological Survey **150** U.S. Geological Survey/photo by T. Miller **151** U.S. Geological Survey **152 (top)** Pierre Guinoiseau/creativecommons.org **(bottom)** U.S. Geological Survey **153 (top)** U.S. Geological Survey/photo by J.D. Griggs **(bottom)** U.S. Geological Survey/photo by J.D. Griggs **156 (top)** University of Colorado, Boulder, CO/National Geophysical Data Center/NOAA **(bottom)** Photographer Howell Williams/National Geophysical Data Center/NOAA **157** U.S. Geological Survey **158** U.S. Geological Survey/Alaska Volcano Observatory/photo by Game McGimsey **159** U.S. Geological Survey/Cascades Volcano Observatory/photo by Lyn Topinka **160** U.S. Geological Survey **162** NASA Earth Observatory image by Robert Simmon, using EO-1 ALI data **166** U.S. Geological Survey/photo by D. Little **167** U.S. Geological Survey/photo by R. McGimsey **168** U.S. Geological Survey/photo by R.D. Miller **169** NASA Goddard Space Flight Center **171** NASA image created by Jesse Allen, using data provided courtesy of NASA/GSFC/METI/ERSDAC/JAROS, and U.S./Japan ASTER Science Team. Image interpretation by Greg Vaughan, Jet Propulsion Laboratory. **172 (top)** U.S. Geological Survey/photo by T.J. Casadeval **(bottom)** U.S. Geological Survey/photo by T. Kobayashi, Univ. Kagoshima **173** U.S. Geological Survey/Cascades Volcano Observatory/photo by Richard P. Hoblitt **174** U.S. Geological Survey/photo by J.D. Griggs **176** NASA Ames Research Center, Tom Trower **178 (top left)** U.S. Geological Survey/Mineral Collection of Brigham Young University Department of Geology, Provo, Utah/photo by Andrew Silver **(top right)** BobMacInnes/creativecommons.org **(bottom)** Wikimedia Commons **181 (top)** Scott Thompson/creativecommons.org **(right)** Library of Congress, Prints & Photographs Division, LC-USZ62-109768 **182 (top)** Felipe Skroski/creativecommons.org **(bottom)** Strange Ones/creativecommons.org **184** U.S. Geological Survey/Cascades Volcano Observatory/photo by Lyn Topinka **194** U.S. Geological Survey/Cascades Volcano Observatory/photo by Rick Hoblitt **195** U.S. Geological Survey/Cascades Volcano Observatory/photo by Mike Doukas **196 (top right)** U.S. Geological Survey/Cascades Volcano Observatory/photo by Peter W. Lipman **(bottom)** U.S. Geological Survey/Cascades Volcano Observatory/photo by Lyn Topinka **197** U.S. Geological Survey/Cascades Volcano Observatory/photo by Lyn Topinka **198** U.S. Geological Survey/Cascades Volcano Observatory/photo by Lyn Topinka **199 (left)** U.S. Geological

Survey/Cascades Volcano Observatory/ photo by Thomas J. Casadevall **(right)** U.S. Geological Survey/Cascades Volcano Observatory/photo by Lyn Topinka **200** K. Segerstrom, U.S. Geological Survey/National Geophysical Data Center/NOAA **201** U.S. Geological Survey **202** U.S. Geological Survey **203** U.S. Geological Survey **204** NASA image by Jeff Schmaltz, MODIS Rapid Response Team **205 (left)** DoD photo by Senior Airman Greg L. Davis, U.S. Air Force **(right)** NASA image by Jeff Schmaltz, MODIS Rapid Response Team at NASA GSFC **207** :mrMark:/creativecommons.org **208** © Terry G. McCrea/Smithsonian Institution